Bradford Monographs in the Archaeology of Southern Asia
1

Agriculture and Pastoralism in the Late Bronze and Iron Age, North West Frontier Province, Pakistan

An integrated study of the archaeological plant and animal remains from rural and urban sites, using modern ethnographic information to develop a model of economic organisation and contact

Ruth Young

BAR International Series 1124
2003

Published in 2016 by
BAR Publishing, Oxford

BAR International Series 1124

Bradford Monographs in the Archaeology of Southern Asia 1
Agriculture and Pastoralism in the Late Bronze and Iron Age, North West Frontier Province, Pakistan

ISBN 978 1 84171 500 1

© R Young and the Publisher 2003

Typesetting and layout: Darko Jerko

The author's moral rights under the 1988 UK Copyright,
Designs and Patents Act are hereby expressly asserted.

All rights reserved. No part of this work may be copied, reproduced, stored,
sold, distributed, scanned, saved in any form of digital format or transmitted
in any form digitally, without the written permission of the Publisher.

BAR Publishing is the trading name of British Archaeological Reports (Oxford) Ltd.
British Archaeological Reports was first incorporated in 1974 to publish the BAR
Series, International and British. In 1992 Hadrian Books Ltd became part of the BAR
group. This volume was originally published by Archaeopress in conjunction with
British Archaeological Reports (Oxford) Ltd / Hadrian Books Ltd, the Series principal
publisher, in 2003. This present volume is published by BAR Publishing, 2016.

Printed in England

BAR titles are available from:

         BAR Publishing
         122 Banbury Rd, Oxford, OX2 7BP, UK
EMAIL   info@barpublishing.com
PHONE +44 (0)1865 310431
FAX   +44 (0)1865 316916
         www.barpublishing.com

Contents

Chapter 1 Introduction 1
1.1   Environmental archaeology in Pakistan and NWFP 2
1.2   Aims and objectives 5
1.3   Defining the study area 5
1.4   Defining the study period 6
1.5   The sites 6

Chapter 2 Methodology 8
2.1   The archaeobotanical and archaeozoological material 8
2.1.1 Taphonomy 8
2.2   Sampling and recovery: collecting the environmental data 9
2.2.1 The Swat Valley sites 9
2.2.2 The Dir Valley sites 10
2.2.3 The Bala Hisar of Charsadda 10
2.3   Identification of the environmental material from the Bala Hisar of Charsadda 10
2.4   Analysing the data 11
2.5   The ethnographic material 12
2.5.1 Collecting the ethnographic material 13
2.5.2 The ethnographic material and continuity 15
2.6   Summary and conclusions 16

Chapter 3 Physical environment and social identity 17
3.1   The Vale of Peshawar 17
3.2   The Northern Valleys 18
3.3   The people 19
3.4   Land use 20
3.5   Factors that may have affected land use 21
3.6   Summary and conclusions 23

Chapter 4 Northern Valleys archaeology 24
4.1   Ghaligai, Swat 26
4.1.1 Interpretation of Ghaligai 26
4.2   Aligrama, Swat 27
4.2.1 Interpretation of Aligrama 28
4.3   Bir-kot-ghundai, Swat 28
4.3.1 Interpretation of Bir-kot-ghundai 29
4.4   Kalako-deray, Swat 29
4.4.1 Interpretation of Kalako-deray 30
4.5   Loebanr III, Swat 30
4.5.1 Interpretation of Loebanr III 31
4.6   Timargarha 1-3 and Balambat, Dir 31
4.6.1 Interpretation of Timargarha 1-3 and Balambat 32

| 4.7 | General interpretation | 33 |
| 4.8 | Summary and conclusions | 34 |

Chapter 5 Southern Plains archaeology 36
5.1	The Bala Hisar of Charsadda	37
5.1.1	Wheeler's chronology	38
5.1.2	The Bradford-Peshawar chronology	39
5.2	Taxila	41
5.3	Summary and conclusions	42

Chapter 6 The Northern Valley sites: environmental material 44
6.1	Ghaligai	46
6.2	Aligrama	47
6.3	Bir-kot-ghundai	48
6.4	Kalako-deray	48
6.5	Loebanr III	50
6.6	Timargarha 1 and Balambat	51
6.7	Environmental material and interpretations from rural Harappan and Bannu Basin sites	52
6.8	Summary and conclusions	53

Chapter 7 The Bala Hisar of Charsadda: new environmental material 55
7.1	The plant material	56
7.1.1	The urban Harappan plant material	57
7.2	The animal bone material	57
7.2.1	The urban Harappan animal bone material	61
7.3	Summary and conclusions	62

Chapter 8 The ethnographic data 64
8.1	The Northern Valleys	64
8.1.1	Winter transhumants	65
8.1.2	Inter-valley transhumants	68
8.1.3	Summer transhumants	68
8.1.4	Nomadic pastoralists	68
8.1.5	Sedentary farmers	69
8.1.6	Northern Valleys: discussion	69
8.2	Charsadda District	70
8.2.1	Charsadda District: discussion	71
8.3	Transhumance in ethnoarchaeology and archaeology	71
8.4	Patterns in the ethnographic record: developing a model	72
8.5	Summary and conclusions	75

Chapter 9 Discussion 76
9.1	Extant theories and models	76
9.2	Urban archaeology and the Southern Plains	78
9.3	The subsistence strategies revealed in the ethnographic material	78
9.4	Testing the new model with the archaeological material	79
9.5	The archaeology and toward a new model	80
9.6	Occupation dates from the Northern Valleys and the Bala Hisar of Charsadda	81
9.7	Implications of the new model: further discussion	81
9.8	Summary and conclusions	82

Chapter 10 Agriculture and pastoralism in the Late Bronze and Iron Age, North West Frontier Province, Pakistan 84

Bibliography 86

Appendices
Appendix 1 Animal Remains from the Bala Hisar of Charsadda 93
    List of bone material: from *in situ* archaeological deposits 93
    List of bone material: by size and element 94
Appendix 2 Plant Remains from the Bala Hisar of Charsadda 98
Appendix 3 Animal Remains from the Northern Valleys 99
    Identified animal bones from Ghaligai, Loebanr II, Bir-kot-ghundai,
        Aligrama & Kalako-deray 104
    List of bone material 104
Appendix 4 Plant Remains from the Northern Valleys 105
    Plant remains from the Northern Valleys 105
    Identified plant remains from Ghaligai, Loebanr III, Bir-kot-ghundai & Aligrama 105
    List of plant material 105
Appendix 5 Ethnographic Interviews 106

List of Tables
4.1   Occupation sequence of Northern Valley sites 24
4.2   Summary chronology of the Northern Valley sites 25
4.3   Northern Valleys sites: pit data 35
5.1   Summary chronology for the Bala Hisar of Charsadda and Taxila 41
6.1   Summary of Environmental material from the Northern Valley sites 44
6.2   Summary of Plant Remains Identified from the Northern Valley sites 45
6.3   Summary of Animal Remains Identified from the Northern Valley sites 45
6.4   Animal remains from Graves: Timargarha 1 51
7.1   Context descriptions for *in situ* archaeological deposits from which the plant
      and animal bone assemblages have been recovered 56
7.2   Plant material from the Bala Hisar of Charsadda 57
7.3   Animal bones from the Bala Hisar of Charsadda 58
7.4   Proportions of sheep/goat to cattle at the Northern Valley sites
      and the Bala Hisar of Charsadda 59
7.5   Representation of skeletal elements (grouped) from the identified bones
      from the Bala Hisar of Charsadda 60
9.1   Summary chronology: ^{14}C dates from the Bala Hisar of Charsadda
      and the Northern Valleys 83

List of Figures
1   Location map North West Frontier Province, Pakistan 2
2   Location map of the sites in this thesis 3
3   Location of ethnographic interviews 14
4   Stylised seasonal movement 65
5   Ground plan of home of Taj Bar, Malam Jabba, Swat 67
6   Ground plan of animal shelter, Malam Jabba, Swat 67
7   Three phase subsistence and settlement 73
8   Sources of influence 74

Acknowledgements

This volume is a revised and updated version of my PhD, submitted to the University of Bradford in September 2000. Many people were involved both directly, and indirectly during the three years of research and writing, and I would like to thank and acknowledge and thank all the institutions and individuals who helped. In particular:

All the academic and technical staff in the Department of Archaeological Sciences, University of Bradford, each of whom at some time has given me help, advice and encouragement.

Professor Duranni, Professor Ihsan Ali, Professor Taj Ali, Mr Mukhtar Duranni, Mr Qazi Naeem, Mr Raj Ali and all the other staff in the Department of Archaeology, University of Peshawar, for their invaluable help and good humour during field seasons in 1997, 1998 and 1999.

The farmers and herders of Charsadda District, Dir and Swat.

Professor Giorgio Stacul, Italian Archaeological Mission in the Swat Valley and University of Trieste, Italy, for making available then unpublished material, and an amazing base of archaeological data.

Alan Braby, Edinburgh, for providing encouragement and transforming rough scribbles into real illustrations.

Amanda Forster, Bradford, for helping with technology.

The other postgraduates at Bradford, particularly Janet Montgomery and Nigel Melton, for discussions, commiseration and support through three years of endurance.

My parents and family for their unstinting support and encouragement.

Janet and Gillian Ambers, for their humour and kindness over many years.

Andrew Reid, University College London, for friendship, advice and support.

My examiners, Julie Bond, University of Bradford, and Robin Dennell, University of Sheffield, for making my viva a humane, even pleasant experience.

Also the staff in the Department of Archaeology and Ancient History, University of Leicester for their help and encouragement, especially, Marilyn Palmer, Graeme Barker, Deirdre O'Sullivan, Jeremy Taylor, Mark Gillings and Dave Edwards.

Finally, thanks to my supervisors Robin Coningham and Jill Thompson, for their complementary skills, advice and help - given unstintingly. Fieldwork with Robin over a number of years and countries, has taught me a great deal and always been interesting. I am grateful for the many opportunities he has given me.

Chapter 1
Introduction

This volume covers the re-assessment of published archaeological environmental material, and the introduction of new archaeological environmental material in order to understand more about subsistence strategies and economic organisation during the mid to late 2nd millennium BC in the north-west of Pakistan. In particular, material from sites in two contrasting regions within modern North West Frontier Province (NWFP) is discussed. First, there are the sites of Aligrama, Bir-kot-ghundai, Ghaligai, Kalako-deray and Loebanr III from the valley of Swat, and Balambat and Timargarha in the valley of Dir. Both of these are steep river valleys, located in the north of the province leading up to the Hindu Kush. Second, to the south of Dir and Swat, within the flat alluvial plain of the Vale of Peshawar is the major urban site of the Bala Hisar of Charsadda. The location of the study area and these sites are shown in Figures 1 and 2.

To date, the main archaeological interpretations have considered these two regions as separate, and this is allegedly supported by differences in social and economic organisation (Stacul 1994a, 1987; Tusa 1979). The valleys of Swat and Dir have been described as both marginal and isolated, and considered closer to Kashmir and China in terms of cultural affiliations than other areas. Indeed Stacul, in support of the 'Inner Asia' cultural concept said that "in the mid-second millennium a cultural frontier crossed the northern hilly area of the subcontinent, including Swat" (1996, 439). This division from, and lack of contact with, the plains area to the south has been supported by the dating of sites in both regions, where ^{14}C estimates from the Northern Valley sites have indicated settlement during PIV (c. 1700-1400 BC) (Stacul 1987) to be much earlier than the dates for the settlement of the Bala Hisar of Charsadda. Chapter 9 considers some of the main theories of cultural development in both the Northern Valleys and in the Vale of Peshawar, and then draws together the results of the analyses of the new and published environmental archaeological data, and the ethnographic data to give a new model.

The earliest settlement dates at the Bala Hisar were, until recently, based on the pottery typology developed by Wheeler (1962) which allowed him to link the development of the site to external, historically recorded events such as the siege of the city of Pushkalavati by Hephaistion in 327 BC. Further, Charsadda had been identified by Cunningham as Pushkalavati, one of the eastern capitals of the Achaemenid Empire (Ali et. al. 1998, 2). The earliest settlement here was thus thought to have been established in the early part of the 6th century BC. However, recent work by the joint Bradford - Peshawar team at the Bala Hisar has resulted in ^{14}C dates indicating that the site was first occupied at least as early as 1400 BC. This means the early occupation is contemporary with the latter part of PIV (according to Stacul's Swat chronology). This period is important in Swat, because it has been identified as discontinuous with preceding periods (Stacul 1969, 56, 62). It is also the period when permanent dwellings are first identified in Swat, and those pits that are assigned a dwelling function are thought to show links with Neolithic Burzahom in Kashmir. Late PIV is believed contemporary with the beginning of the Gandharan Grave Culture in this area (Stacul 1966, 84).

Analysing and interpreting the archaeological environmental data from the Northern Valley sites and the Bala Hisar of Charsadda gives the opportunity to compare results from these two contrasting environmental areas, and from urban and rural sites. The introduction of other archaeological data allows a re-evaluation of the theories and models of social and economic organisation and change in each region. When the results of the ethnographic interviews are introduced, a new model can be developed from this material and tested using the archaeological data. The results of the ethnographic interviews, which are largely corroborated by historical information, indicate that rather than the simple plains/hills and rural/urban dichotomy underlying the extant models of subsistence and economic organisation, there is in fact a complex subsistence mosaic.

In order to place this work in context it is necessary to summarise and critically assess the current state of environmental archaeology within Pakistan, and more particularly NWFP. This is important because the current

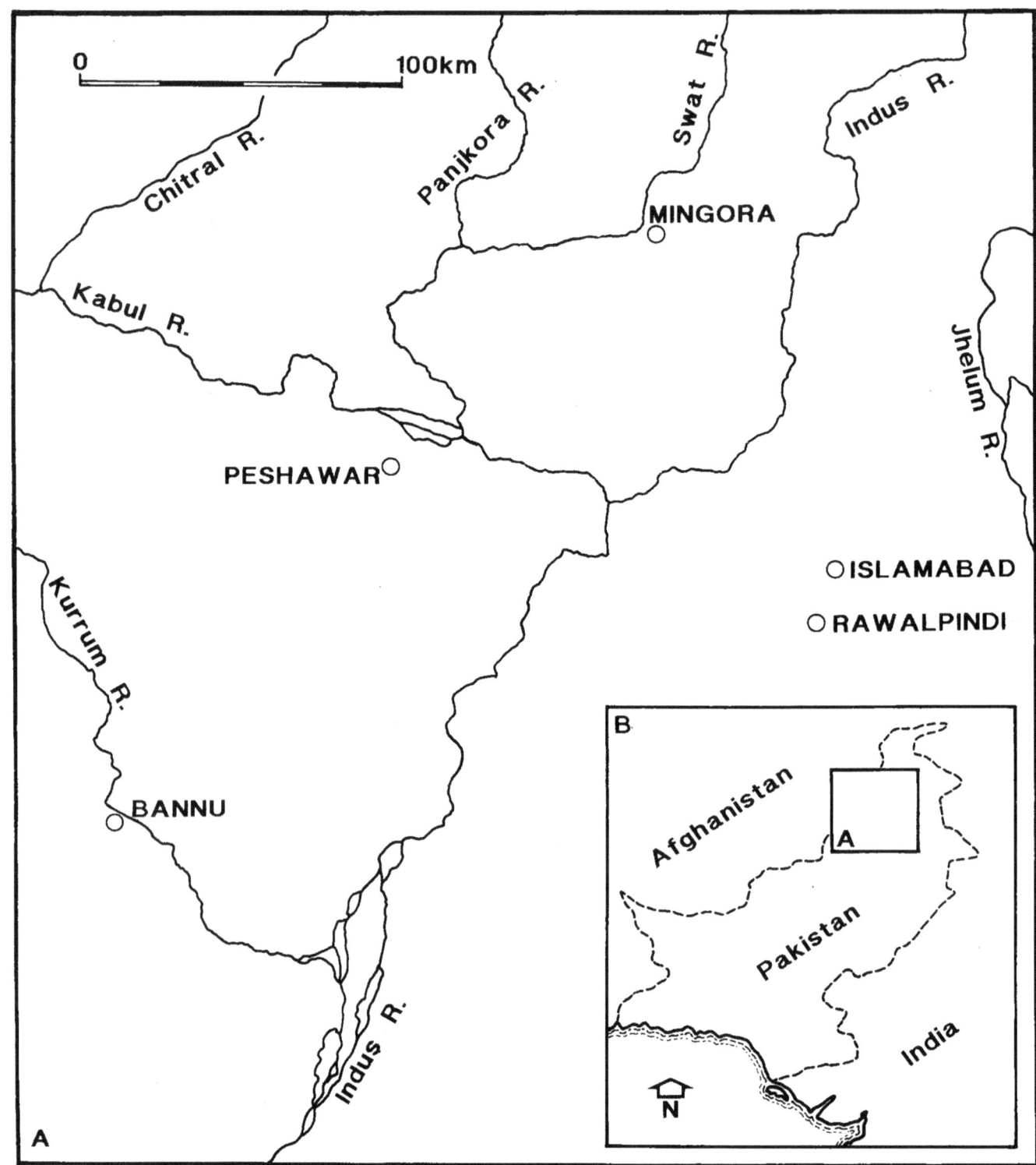

Figure 1. Location map of North West Frontier Province, Pakistan (Alan Braby 2000)

status and use of environmental archaeology in this region are crucial to understanding why this reasearch has taken this particular form, and further, why certain methodological approaches have been used. A definition of the study area and the nomenclature used in this volume in relation to the specific areas of research will be given, and then an explanation of why the sites in each region have been selected for study here. The final section of this chapter will explain why the terms 'Bronze Age' and 'Iron Age' have been retained, and a definition appropriate to their use in this work given.

1.1 Environmental archaeology in Pakistan and NWFP

Over ten years ago Meadow said of archaeozoological research in Pakistan that "to date, it has been extremely difficult to do more than generally characterise the situation in the Greater Indus Valley as a whole and to use analogy with modern conditions as a guide to the past" (1989, 71). Since then, there have been a number of specialist reports examining very specific areas of plant and animal

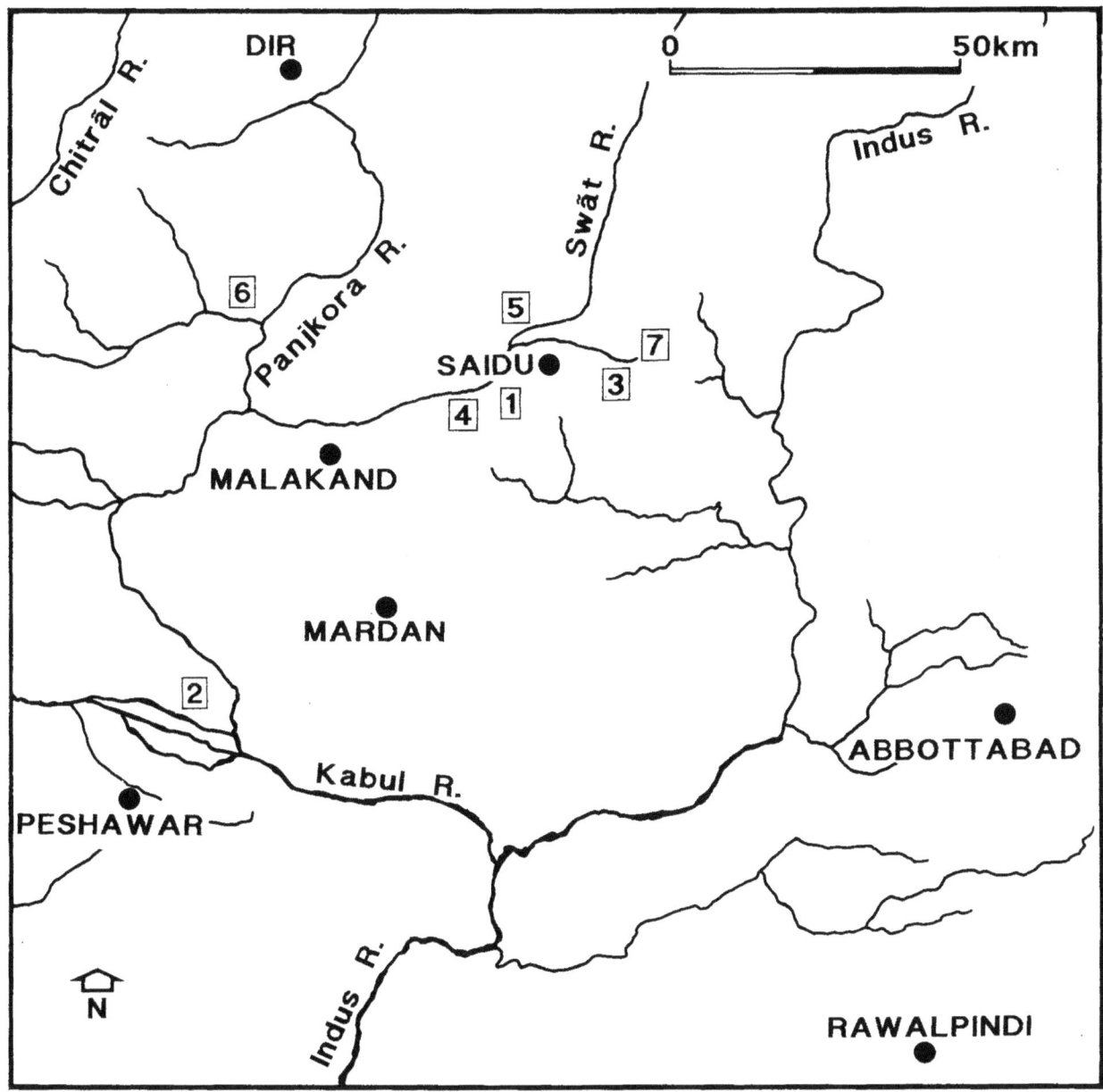

Figure 2. Location map of the sites in this book: 1. Bir-Kot-Ghundai; The Bala Hisar of Charsadda; 3. Loebanr III; 4. Ghaligai; 5. Aligrama; 6. Timargarha & Balambat; 7. Kalako-Deray (Alan Braby 2000)

exploitation. Faunal studies include Belcher's examination of fishing strategies (1994), Meadow's study of the possible correlation between bone measurements and animal husbandry practices (1991) and Kenoyer's study of shell working (1991). Botanical research includes Weber's palaeoethnobotanical work aimed at understanding plant use and agriculture in the Indus (1991, 171), and more particularly food stress as an explanation for culture change (1999; 1992, 258). Miller has presented a preliminary investigation of urban palaeoethnobotany from Harappa (1991).

These examples however, are all from Harappan sites that have been excavated over many years, using scientific methods of data collection, thus providing a large body of suitable data. The general archaeological focus within Pakistan has centred largely on Harappan sites and Gandharan art, so it is unsurprising that the vast majority of environmental work centres on sites associated with the former as well. Possible reasons for this situation include the culture-historical legacy of Wheeler, one of the most influential figures in the history of archaeology in Pakistan. In his summary of the history of archaeobotanical research in India, Fuller says that the earliest phase of research (pre Independence until 1974) was characterised by archaeologists sending any recovered plant material to botanists for identification. Further, "the material recovered was generally small in quantity and collected unsystematically, whenever an excavator noticed a quantity of recognisable plant material, such as a ceramic vessel full of charred grain, or fortuitously preserved pieces of wood" (1999, ms. 3).

With the exception of the Harappan sites as noted above, this is still largely the situation of published work in Pakistan today. Of nine non-Harappan sites in Pakistan from which plant remains have been recovered and published, only three have had flotation carried out. One of these, Tarakai Qila (Thomas 1999) dated to the 3rd millennium BC, will be discussed further in Chapter 6. At the Makran site of Miri

Qalat, and Shahi-Tump in Baluchistan, where occupation has been dated to 4000-2000 BC, placing it in the proto-historical period, botanical samples have been taken from a range of contexts, and plant macroremains recovered by soil flotation (Tengberg 1999, 3-12).

The work by Thomas (1999, 1986, 1983) originally as a member of the British Archaeological Mission to Pakistan is an exception to this overall situation. This project, which is an exploration of the archaeology of the Bannu Basin, included within its research design the express aims of sampling for subsistence remains, and evidence for past environmental conditions, with the subsequent analysis and interpretation an integral element (Thomas 1983, 13). The results of fieldwork from sites spanning over 2000 years of occupation, from the Early Chalcolithic Sheri Khan Tarakai (c. 4240 BC) to the Early Harappan Tarakai Qila (c. 2042 BC), have allowed the development of a model of subsistence and change in the Bannu area (Thomas 1999, 312) .

Work in Swat has been undertaken by members of the Italian Archaeological Mission of IsMEO (*Istituto Italiano per il Medio ed Estremo Oriente*) which is now the Italian Archaeological Mission of IsIAO (*Istituo Italiano per L'Africa e L'Oriente*). This major project was begun in the 1950s (Facenna 1964), and continues today with exploration at Bir-kot-ghundai and recent work at Kalako-deray (Stacul 1997b). A limited amount of environmental material has been recovered, identified and analysed from the excavations, and this is discussed further in Chapters 2 and 6.

To understand why there is a relatively small amount of environmental data collected from NWFP (with the exception of the work by Thomas et al. in the Bannu Basin), it is necessary to look at the research designs and questions that are the driving force behind the excavations conducted in this region. At the sites of Ghaligai (Stacul 1967a) and Loebanr III (Stacul 1976) in Swat, excavations were carried out with the primary objective of learning about the pre-Buddhist cultural and chronological sequence, mainly through the structural and artefactual remains. In Dir, the excavations at Chakdara and surrounding sites were aimed at building a cultural sequence and chronological time scale, and also to record the art and artefacts uncovered (Dani 1968-9, 6-7). At Timargarha and Balambat, also in Dir, the excavation and study of the graves and artefacts were aimed at defining the cultural group known as the Gandharan Grave people (Dani 1967, 24-30). In his analysis of the excavated material from Dir, Dani said "Pottery is the most important find in the graves at Thana and Timargarha" (ibid., 121).

With regard to sites in the Vale of Peshawar and further south in NWFP, research designs were aimed at understanding urban development. The most recent excavations at the Bala Hisar of Charsadda (Ali et al. 1998) were guided by a research design based on seven main questions. All of these questions were concerned with clarifying the chronology of the site, and shedding new light on the claims Wheeler (1962) made following his excavations here in 1958. In particular the date and course of the defensive ditch and the relation between the ditch and other features, the walling of the eastern mound and the nature of the prehistoric river channel between the eastern and western mounds were being investigated. At the site of Hund situated on the banks of the Indus, excavations were aimed at providing corroboration of the historical accounts that attest to the greatness of this city through many centuries (Ali, n.d., 3). Rehman Dheri in the Dhera Ismail Khan District, is one of the earliest known urban sites in this region, with a long multi-period history. One of the excavation goals is explicitly stated as an opportunity to "collect botanical and faunal remains that would have some bearing on the economy and ecology of Rehman Dheri through time" (Durrani 1988, 19). However, within the excavation report there is no account of an environmental sampling strategy, nor are any preliminary results given.

Many Buddhist sites from the Historic Period in NWFP have been studied, but have almost completely ignored environmental questions and sampling. For example, in the recent excavations at Pir Manek Rai in the Haripur Valley, excavated by a joint team from the Federal Department of Archaeology of Pakistan and the National Heritage Foundation (Durrani et al. 1997, 216), the location of the site, and role of natural resources in its settlement history and cultural development are stressed. Yet there was no sampling strategy for plant remains, and a very few hand collected animal bones were made available for recording and identification (Young 1997).

This brief examination of research designs for some of the excavations that have provided environmental data from non-Harappan sites, shows that the questions being asked are more concerned with issues of chronology, culture history, population contact and movement rather than environmental exploitation and subsistence. When attention has been given to subsistence, assumptions have been made on the basis of such remains as pottery types, or from a cursory analysis of small amounts of data. For example, Stacul and Tusa reported finding very little in the way of animal bones from the excavations at Aligrama in Swat during the 1962 and 1975 seasons (1975, 295). However, they did recover many miniature vases, and said of these "This particular type of utensil must be connected with the necessity of a people who lived mainly by stock-raising for preserving small but precious substances useful and indispensable for fermenting milk and forming curds for making milk products" (ibid., 309).

In summary, it can be seen that within the archaeological work carried out in NWFP, only a very small proportion of the excavations have made environmental or subsistence questions either a priority or even an integral part of the research designs. This is because the research designs for the projects carried out in both the Swat and Dir valleys, and in the Vale of Peshawar and other parts of lowland NWFP are clearly aimed at exploring other issues. The prioritisation of these issues is no doubt shaped by the overall demands of archaeological research within both NWFP and Pakistan. The development of archaeology here occurred within the culture-historical paradigm introduced and developed by Marshall and Wheeler, among others. Unlike neighbouring India, where the change in the underlying archaeological framework following Independence has led to a much greater development of environmental archaeology (Fuller 1999 m.s. 6), for example at Daimabad (Vishnu-Mittre et al. 1986, 623-7), the focus within Pakistan remains on chronology and pottery typologies.

1.2 Aims and objectives

The aim of this volume is to develop a greater understanding of the subsistence strategies of the occupants of two environmentally contrasting areas of NWFP during the Late Bronze and Early Iron Age and to develop a new explanatory and predictive model to relate the environmental conclusions to these subsistence strategies in these regions. The two physical areas are the Charsadda District in the Vale of Peshawar, and the valleys of Dir and Swat to the north (see Figure 1.1) and according to interpretations of the archaeology by the excavators, and other scholars, the sites in these two areas represent very different cultural groupings (Allchin 1995, 30; Dani 1968, 1992; Stacul 1987; Tusa 1979; Wheeler 1962).

This aim is achieved through a number of objectives. The first is the analysis and interpretation of archaeological environmental material from sites in both these areas. The second objective is to determine whether urban and rural sites can be distinguished on the basis of their faunal and floral assemblages. Third, environmental and other archaeological data will be used to test selected existing models and theories of subsistence and development within each of the study regions, and also contact between these regions. Further objectives include determining whether in this region the archaeological environmental material provides any information about mobile subsistence strategies practiced by groups in either areas, and how these can be distinguished from sedentary groups. In terms of testing existing theories and models of subsistence in the study area, this is carried out using both environmental and other archaeological data, and a critical assessment of how far the data supports the theories made.

The settlement period of interest in this study is between c. 1700-1000 BC according to the individual site, and this period between the end of the Harappan Civilisation and the emergence of the Early Historic Cities at Taxila and Charsadda, traditionally assigned to c. 600 BC (eg Dani 1992, 395; Marshall 1951, 11; Wheeler 1962, 34) has been described as one of cultural discontinuity (see Coningham 1995 for opposite argument). Reappraisal of the archaeological and environmental material, in particular new information with regard to dates of establishment of these sites, will allow testing of the models of discontinuity and external influence in both study regions.

The above account of the priority given to environmental archaeology in NWFP makes clear the need to demonstrate the value of sampling, analysing and interpreting plant and animal remains in terms of understanding more about development and change at archaeological sites. As will be shown in Chapters 2, 6 and 7 (and in Appendices 1-4), only very small assemblages have been available for study. This has resulted in the integration of both plant and animal remains, which has allowed a more complete discussion of subsistence strategies. In addition to the archaeological environmental material, a series of ethnographic interviews were carried out over two field seasons, the purpose being to develop a new model of land use and subsistence in the study area with which to test the archaeological data. In addition, utilising new ethnographic data means that it is possible to begin to determine what factors are significant in shaping subsistence strategies within the different study areas. From this, an assessment of how important the geographic setting and attendant environmental constraints are in these choices, and how significant other issues such as ethnicity and ideology are, can be made.

1.3 Defining the study area

An early Gazetteer (Imperial Gazetteer of India 1904, 1) divides NWFP into three main geographical areas. They are: 1, Hazara, or the cis-Indus region; 2, the strip between the Indus and the hills to the west, comprising Peshawar basin, Kohat and the Bannu-Dera Ismail Khan strip; 3, the mountainous regions to the north and west of the first two. Figure 1.1 shows the location of NWFP within Pakistan, and Figure 1.2 shows the main sites mentioned in this work. The following is an outline of the terms used within in this volume, and the areas, sites and groups of sites that are referred to.

Area: this refers to the total study area, which is the area of the Vale of Peshawar, and so including the archaeological site of Charsadda, and the two valleys of Dir and Swat, both located to the north of the Vale of Peshawar. The sites of Balambat and Timarghara are located in Dir, and those of Aligrama, Bir-kot-ghundai, Ghaligai, Kalako-deray and Loebanr III are all in Swat.

Region: each of the two distinct geographical components of this study is referred to as a region. Therefore, the valleys of Swat and Dir comprise one region, and the Charsadda District another. Where the site of Taxila is used in the discussion, as a site comparable with Charsadda, it is included in the general designation of southern or plains region.

Northern Valleys: the specific region of Swat and Dir, two river valleys to the north of the Vale of Peshawar.

Southern Lowlands or Plains: the specific region encompassing the Vale of Peshawar, but not confined to it. When appropriate, this also includes the region around Taxila, which falls in the modern province of Punjab.

Vale of Peshawar: the distinct geographical area in the shape of a basin, being a flat plain surrounded on three sides by significant hill ranges, and in the east at the confluence of the Kabul and Indus Rivers, open towards Taxila.

Charsadda District: for the purposes of this study, Charsadda District is the area of modern towns and villages, and farms surrounding the archaeological sites of the Bala Hisar of Charsadda and the eastern mound. Charsdda District is in the Vale of Peshawar.

Dir: the valley of Dir lies to the north west of the Vale of Peshawar, and is also a modern political district, with its boundaries following the natural barriers of hills.

Swat: the valley of Swat lies to the north east of the Vale of Peshawar, and like Dir, the whole valley comprises a modern political division, based on natural barriers.

While some of these terms equate to modern definitions, and others to accepted text-book descriptions (eg Dichter 1967, 32, 52, 91-2), they are on the whole specific to this study, and throughout the study are used in accordance with the above outline.

1.4 Defining the study period

The title of this volume refers to the Late Bronze Age and Iron Age with regard to the chronological period under study. It is recognised that it is not appropriate to transfer directly the established European concepts and definitions of Bronze and Iron Age, with their implicit assumptions of cultural packages and linear development. However, within this work, and within the wider framework of archaeological interpretation in Pakistan, the use of Bronze and Iron is intended to conform with the accepted description of the appearance and domination of these types of metal artefacts on sites.

According to the excavation reports from the study area (eg Dani 1968; Stacul 1987; Wheeler 1962), there is a progression in terms of metal artefacts from bronze at earlier sites, and in earlier levels, followed by iron at a later stage. While the appearance of iron has been attributed variously to incoming groups such as Indo-Europeans or Achaemenids, and so been made a vehicle for diffusionist, if not invasion theories, on archaeological grounds it certainly appears to follow bronze in this region. Given the rapid and significant changes to the chronometric dates obtained in this area (see Chapters 4 and 5), to assign absolute dates to each region, or to the area as a whole for the period being studied would open up other difficulties. Primarily, these dates would almost certainly be 'outdated' following new excavations that made sampling and measuring for chronometric dates a research objective. Rather, as Wheeler noted of his own work on the chronology at the Bala Hisar of Charsadda which linked the pottery types from the three trenches into a single dating scheme "the whole of this tentative time-table cross-checks rationally and is unlikely to be far wrong" (1962, 36), and while the absolute dates have changed, Wheeler's internal typology has not been challenged. A further reason for using these relative terms is that they are used and recognised by Pakistani archaeologists and archaeologists working in Pakistan, and so can easily be related to other work within both NWFP and Pakistan.

Although some radiocarbon dates are available for many of the sites from the Northern Valleys in addition to those from the Bala Hisar of Charsadda, there are still some problems comparing these. The dates from the Northern Valleys have been calibrated (or even re-calibrated) by Possehl (1994), but there are at most two estimates for each site. Further, it has not always been made explicit what material has been used for the samples, or from what archaeological contexts they have been derived. Only from the Bala Hisar is such information consistently available (Ali et al. 1998; Coningham pers. comm., 2000).

1.5 The sites

Archaeological exploration in Swat, Dir and other adjacent regions such as Buner, during the 1950s (Dani 1967; Facenna 1964; Stacul 1967b) resulted in the discovery of a number of grave sites with similar construction and pottery. These graves have been attributed to a cultural group, variously named the Gandharan Grave people (Dani 1968, 24) or pre-Buddhist (Stacul 1966, 37), and further work uncovered a number of settlement sites associated with some of the grave sites, but also pre-dating them. Gandharan Graves have since been noted in the Vale of Peshawar and Punjab (Allchin 1993; Khan 1993), but the connection between these sites and the urban sites of Charsadda and Taxila has not been fully explored. It has been suggested that there was no connection between the urban sites and those of the Gandharan Grave people (Dani 1992, 395-6). This study does not examine the graves, or material from the graves (with the exception of material from Timargarha 1 in Dir), but looks at material recovered from occupation sites, which may be associated with Gandharan Graves.

The selection of the sites used in this study was necessarily limited to those that had new environmental material available for analysis, or that had environmental material published. New material was available from the Bala Hisar of Charsadda, and this is listed in full in Appendices 1 and 2, and analysed in Chapter 7. Published material was available for the sites of Aligrama, Bir-kot-ghundai, Ghaligai, Kalako-deray and Loebanr III from Swat, and a note about animal bones recovered from the graves at Timargarha in Dir had been published. The Swat material is listed in Appendices 3 and 4, and along with the Timargarha material is analysed in Chapter 6. The sites from Swat are all occupation sites, and it was considered important to study these rather than the material from the grave sites to gain greater understanding of the subsistence strategies. The grave material from Timargarha 1, rather than Balambat, the Dir settlement site was used simply because this is the only published data available for Dir.

There are many problems involved in the analysis of small environmental samples, and those collected according to different strategies for different aims, and these problems are compounded by inadequate information about their archaeological contexts and phases. Despite these problems, the material still exists, and it is considered important to utilise what is available. The recovery of both plant and animal remains from a number of sites in the study area indicates that preservation conditions for both are adequate, and so comprehensive sampling and recovery programmes would probably result in significant assemblages. However, dedicating time and labour to such recovery, identification and analysis will only occur if it can be demonstrated that environmental material is as important in the understanding of past site occupation and economic organisation as pottery in South Asia.

Chapter 2
Methodology

This is an outline of the methodology relevant to the two main areas of this research; firstly, the handling of both new and published archaeobotanical and archaeozoological data, and secondly, the collection of new data in the form of ethnographic material, and will cover the sampling and interviews.

2.1 The archaeobotanical and archaeozoological material

Some of the main recognised taphonomic processes that may affect the composition of plant and animal assemblages, and in particular, those that have played a role in the analysis of material from South Asia, will be summarised. This will be followed by an outline of the environmental material from the sites, and the sampling and recovery methods used by the excavators and specialists at each. With regard to the new material from the Bala Hisar of Charsadda, the identification techniques for both the plant and animal assemblages will be noted, and then analytical techniques appropriate to both small assemblages, and those collected and identified according to different criteria will be discussed.

2.1.1 Taphonomy

There are many potential sources of bias in environmental assemblages and a great deal of literature is devoted to investigating and allowing for these biases during analysis. Taphonomy is the study of the processes affecting organic materials as they become part of the archaeological record (Davis 1987, 17), or "the study of the transition...of organics from the biosphere into the lithosphere of geological record" (Lyman 1994, 1).

Plant remains are preserved by a number of means, including mineralisation, waterlogging and desiccation, but by far the most common means of preservation, certainly in South Asia, is that of charring. According to Hillman "on most well drained sites...plant materials are preserved by the chance exposure to fire and, even then, only when heating is relatively gentle (200-400°C) or, if temperatures are higher, when they are smothered in the ashes...such that they are preserved intact by charring rather than being burned away altogether to mineral ash" (1981, 139). However, once plant materials have been charred and reduced to an estimated 50 to 60% elemental carbon, they are considered to be resistant to organic decay, and likely to be damaged only by mechanical activity (Miksicek 1987, 219).

As part of the conclusions of his study of plant macroremains assemblages from three Bulgarian sites, Dennell (1972, 157) emphasised that when extensive sampling and recovery programmes were implemented, there could be a significant variation in the range and abundance of taxon identified from each sample. This study demonstrated that the contexts and sources of the assemblage had a clear effect on the results, and for example, could lead to biases in the interpretation of the importance of some plants compared to others (ibid., 158). This needs to be taken into account during analysis and interpretation, especially where samples are small, and material from different types of contexts are being used.

Animal bones and teeth, and less frequently horn and antler, will survive in a range of environmental conditions, and this survival is related to the physical properties of different skeletal elements themselves (Lyman 1994, 72-81; Reitz & Wing 1999, 118). The physical conditions of the site may also be important, for example, the type of soil is thought to be a crucial factor in preservation. It is generally considered that open sites with a neutral to alkaline pH reading are most favourable to bone preservation (Cresser et al. 1993, 70; Miksicek 1987, 214). However, preliminary investigation on a northern British site comparing bone preservation to sediment pH suggests this may not be the simple correlation previously asserted (Young & Nicholson 1998-9). A wide range of biological and chemical processes may be at work on bone after deposition and prior to recovery. Weathering, insect and rodent activity, and the leaching of salts from surrounding soil are just some of the possible processes, and these are summarised by O'Connor in his Table outlining the 'Subdivisions of the taphonomic process' (O'Connor 2000, 20). Scavenging could also include scavenging by birds, and this could result in the removal of bones some distance from a site (Bond & O'Connor 1999, 420). Excavation and recovery techniques have also been shown to have a major effect on the composition of a faunal assemblage. A strong bias toward the recovery of large bones, evident in large animals being represented in a site's species profile, has been shown as a result of hand excavation only (Davis 1987, 29-31), and soil sieving clearly results in the recovery of greater numbers of small and delicate bones from a range of species.

Like plant remains, the animal bones recovered from any given context on any given site are usually present as a result of a number of deliberate or unconscious selection processes. Certain animals are often perceived as having a symbolic as well as a practical role within many societies. Shaffer and Lichtenstein argue that cattle represented cultural wealth during the Harappan period in South Asia (1995, 144-5). Using the example of Oriyo Timbo, a Harappan site in India, they claim that as agriculture developed, the problem of keeping herds away from crops increased. This apparently led to the creation of seasonally specialised cattle settlements, which in turn led to cattle being seen as a focus of wealth and their handlers or herders as specialists, important within the developing occupation and kinship ties of the Harappan. Similarly, following a study of faunal material from sites in southern and eastern Africa, Reid says that "animal remains from the sites of complex societies are, therefore, not just artefacts of power relations over allocated resources of cattle, but are also manifestations of relations regarding authoritative resources" (1996, 46). Just as the analysis of plant remains in terms of crop processing has been used to try to address questions of economic organisation, these examples show how faunal data can be used to help explain social and political activities.

2.2 Sampling and recovery: collecting the environmental data

For the purposes of this section, the sites will be separated thus: the sites in the Swat Valley, which were all excavated under the auspices of the Italian Archaeological Mission to Pakistan; the sites in the Dir Valley, which were excavated by a team from the University of Peshawar; and the Bala Hisar of Charsadda, which was excavated by a joint team from the Universities of Bradford and Peshawar. This division has been made to cover the three different approaches to environmental data sampling and recovery.

2.2.1 The Swat Valley sites

The sites in Swat that have published animal remains are Aligrama (Compagnoni 1979), Bir-kot-ghundai, Ghaligai, Loebanr III (Compagnoni 1987), and Kalako-deray (Jawad 1998), and those that have published plant remains are also Aligrama (Costantini 1979) Bir-kot-ghundai, Ghaligai and Loebanr III (Costantini 1987). Table 6.1 (Chapter 6) summarises the environmental material recovered from these sites, and the methods used. These sites were variously dug between 1956 (Stacul 1987, 19) and 1997 (Stacul 1997b, 363) under the direction of the Italian Mission of IsMEO (*Istituto Italiano per il Medio ed Estremo Oriente*), which is now the Italian Archaeological Mission of IsIAO (*Istituto Italiano per L'Africa e L'Oreinte*).

The animal remains from Aligrama, Bir-kot-ghundai, Ghaligai and Loebanr III were identified and analysed by Bruno Compagnoni of the *Servizio Geologica d'Italia* in Rome (Compagnoni, 1987, 1979), working in collaboration with Lucia Caloi on the early Loebanr III material (Caloi & Compagnoni 1976). The animal bones from five pits at Kalako-deray were identified and analysed by Alia Jawad of the University of Peshawar, working in conjunction with IsMEO (Jawad 1998). The botanical remains from Aligrama, Bir-kot-ghundai, Ghaligai and Loebanr III were all identified and analysed by Lorenzo Costantini working in conjunction with IsMEO (Costantini 1987, 1979). No information is given in the published reports about the methods of collection of the animal bones from Ghaligai, Loebanr III or Bir-kot-ghundai. Likewise, the report on the bone material from the pits at Kalako-deray does not give specific details of how the assemblage was collected. In the absence of details of sieving of deposit from any of these sites, it is assumed that all faunal material identified and analysed was collected by hand excavation.

With regard to the sampling and recovery of plant remains, at the site of Ghaligai, Costantini says that "sampling for palaeoethnobotanical analyses was rather random" (1987, 155). He clarifies this, by stating that samples were collected opportunistically from hearths or other features that were thought to contain plant materials and large fruit stones were collected by hand when recognised (ibid.). At Bir-kot-ghundai "some soil samples were taken for paleoethnobotanical analysis. They come from the earliest occupation layers of the dwelling" (ibid., 156). No details of the volume of the samples or their treatment are given. The excavation of the site of Loebanr III uncovered a large number of pits of various sizes and shapes, and samples for palaeoethnobotanical analysis were taken from the fill of these pits. Costantini describes the sampling here as "a biased palaeoethnobotanical study, based on an analysis of separate samples that were selected for various reasons, and which can all be dated to Period IV" (ibid., 158). All the material included in the analysis was preserved by charring. Some plant material recovered from Pit H1 at Loebanr III was uncharred, but was discarded by Costantini as the likely result of modern intrusion (ibid., 158). For Aligrama and the early investigations at Loebanr III, Compagnoni says the "soil samples were gathered during the course of the excavations without following any predetermined research system but rather adapting each time to the varying archaeological conditions. About half of each sample was washed *in situ*" (Compagnoni 1979, 703). No details are given about the size of each sample.

2.2.2 The Dir Valley sites

The grave site of Timargarha and the settlement site of Balambat in Dir were excavated between 1964 and 1966 by a team from the Department of Archaeology, University of Peshawar, led by Professor A. H. Dani (Dani 1967). The animal bones recovered from Timargarha were identified and analysed by Dr Hemmer, a zoologist at the University of Mainz (Bernhard, 1967, 370). Animal bones were noted from the settlement site of Balambat during excavation (Dani, 1967, 242, 243) but there is no discussion of them at all. No sampling for plant remains was carried out at either site, and other than the mention of charcoal in occupation layers, for example at Balambat where charcoal mixed with loose soil and ash is described as part of an occupation phase (ibid., 244), no discussion of plant remains occurs in the reports on these sites.

The animal remains recovered from the site of Timargarha were collected by hand as part of the excavation of the graves (Dani 1967, 31). These remains were considered important in so far as they were recovered in conjunction with human remains and so "are furthermore of special archaeological and ethnological interest because they seem closely connected with certain ritual practices" (Bernhard, 1967, 370). There has been no attempt to analysis the animal remains from either site in terms of the subsistence strategies of the occupants.

2.2.3 The Bala Hisar of Charsadda

The Bala Hisar at Charsadda was excavated between 1994 and 1996, by a joint team from the Universities of Bradford, England and Peshawar, Pakistan and was directed by Dr Robin Coningham, and Professors Abdur Rehman, Taj Ali and Ihsan Ali. The aims of the excavation seasons at the Bala Hisar of Charsadda are discussed elsewhere (see Chapter 1 and Chapter 5), and these aims and other constraints limited the environmental investigation undertaken at this site. According to the Field Director, "bearing in mind these defined aims and objectives, we were unable to dedicate substantial resources or personnel to the collection of environmental material. Indeed, due to the restricted nature of our trenches, in many cases we were unable even to ascertain the function of many of the archaeological deposits encountered" (Coningham, pers. comm., 2000). The absence of an environmental sampling strategy has resulted in very small assemblages that contain many inherent biases, and these biases and their implications are discussed further below. The faunal and floral material that was recovered was processed and analysed by the author; the faunal material was examined in Peshawar, and the floral material in Bradford.

The faunal material from the Bala Hisar at Charsadda was collected by hand during excavation when recognised, and through retention on the 4mm sieve that was used to screen all the deposit excavated on site (Coningham, pers. comm., 1999). The condition of the bone ranged from extremely good to very poor. The poor bone was both crumbling and fragmentary, making recording and identification difficult.

The majority of plant macroremains samples collected at Charsadda were intended for ^{14}C dating purposes and comprise wood charcoal pieces. Plant macroremains other than wood charcoal derive from two sources. The first is hand-collected clumps of burnt grain that were clearly visible in the soil, and thus recognised and excavated. The remaining samples were taken when charred material was visible in less intense quantities, but nevertheless still recognisable, and this was achieved by collecting approximately two litres of the soil matrix from occupation layers, within pots and other specific areas. The soil samples were then wet sieved in the laboratory using a 500-micron sieve, and the resulting residue sorted by hand under a microscope. All the plant remains recovered and discussed here were preserved as a result of charring.

It is clear that the recovery of the environmental material available to explore questions of subsistence in this research was flawed. All the samples are small and the sampling and collection program at each site was unsystematic. However, environmental material does exist, and it is considered important that an analysis and subsequent interpretation of it is carried out. Ignoring it, or simply listing it as an appendix to the main excavation report detailing architecture, culture sequences and pottery typologies will not advance the cause of environmental archaeology in this region. Through asking simple questions of the analysed data, some insight into subsistence patterns may be gained, and recognition of changes may be possible. Integration of different types of environmental data gives a more complete subsistence picture, and synthesising data from a number of sites can be useful in detecting trends where samples are small. For example, Halstead's (1992, 22) study of the faunal remains from Neolithic sites in Greece drew on samples ranging from 141 identified specimens from one site, up to 2256 identified specimens from another. Using the analysed environmental material in conjunction with other archaeological data, and also the ethnographic material relating to subsistence patterns in the area of study, a model can be developed that can begin to address some of the questions regarding the economic organisation and archaeology of the area.

2.3 Identification of the environmental material from the Bala Hisar of Charsadda

The botanical data from the Bala Hisar of Charsadda was identified using a combination of published and reference material, based on the examination of the external morphology of the charred plant macroremains. The reference works used included Zohary and Hopf (1988) for cereals, Thompson (1996) for rice, and Schoch et. al. (1988) for weed seeds. The plant material recovered and identified is listed in Appendix 2.

The animal remains from the Bala Hisar of Charsadda were identified using a combination of comparison with literature sources, and modern reference specimens. The archaeological animal remains could not be removed from Pakistan, and so were examined in the Department of Archaeology at the University of Peshawar. The main reference works used were Hillson (1992) and Davis (1987), with Higham (1975) particularly useful in helping distinguish between cattle and buffalo remains. The Department of Zoology, University of Peshawar allowed access to their reference material and gave advice.

The majority of the Charsadda animal bone assemblage was in good condition, although some specimens had sustained modern damage from the excavation process. Some of the bone was in poor condition, and regardless of condition most of the bone was quite fragmented. A small proportion of the assemblage showed signs of having been exposed to fire prior to burial, and this was recorded along with butchery marks. Age estimates were based on epiphyseal fusion of post-cranial bones using the scheme from Silver (1969) as a general guide. Dental eruption and wear were recorded according to the scheme used by Payne (1973), supplemented by Hillson (1986). Where possible, adult post-cranial bones were measured, and assigned to body side.

Given time and other pressures while identifying this assemblage, and the research questions, it was not considered a worthwhile investment of time to differentiate (where possible) between sheep and goat. This was also compounded by the small size of the assemblage, which allowed little internal differentiation and comparison. Sheep and goat are therefore referred to together as caprine. This is consistent with the other archaeozoological reports used in this study, as Compagnoni (1987) has united sheep and goats in the report from the three Swat faunal assemblages, and Jawad (1998) also grouped sheep and goat as *Ovis/Capra*.

2.4 Analysing the data

The objective of analysing the environmental material was to recognise trends within and between the data sets. Analysis of plant and animal assemblages should allow the development of models in order to make predictions suggesting what data could be expected from different subsistence approaches within the archaeological material, and how these might change over time (Popper 1988, 58). To obtain the 'best' analysis of the available data, thus being able to work towards the most comprehensive and best informed interpretation, it is necessary to select the most appropriate means of analysis, bearing in mind both the research questions being asked and the data (Bond 1994, 140; Hubbard and Clapham 1992, 117-8; Popper 1988, 53-4). There are two types of analysis: qualitative and quantitative, and these will be discussed, with those methods considered suitable for application to the data used in this work noted.

In the study area (the Vale of Peshawar and the Northern Valleys of Swat and Dir) where very little extensive environmental archaeological study has been either carried out or published, this is the first attempt to integrate more than one source of environmental data. By examining the characteristics of the subsistence patterns in the sites from the Northern Valleys and those from the Vale of Peshawar and surrounds, changes over time should be evident in the data, if the archaeological contexts and dating allow such distinctions. Therefore, an approach that begins with what Pearsall calls 'simple tabulation' (1989, 196), which will highlight main trends in the data, and form the basis of qualitative analysis is appropriate here. Once the material has been recorded and presented in such a way that allows basic comparisons between and within sites, this may encourage the collection of better data sets. This rests on better sampling procedures leading to larger samples and identifiable specimens, and these in turn will allow more detailed questions to be asked and the application of more sophisticated analytical techniques.

A number of quantitative analytical techniques have been applied to environmental data. However, even with data sets far larger, and more systematically collected than those forming the basis of this study, there are a number of criticisms that have been levelled at such analysis. Hubbard and Clapham (1992, 119), for example, maintain that detailed descriptions and sophisticated analysis of plant assemblages is only really worthwhile if the samples are taken from well defined and absolutely unambiguous contexts. Generally, the majority of plant remains from any site are more likely to be from pit or ditch fills where mixing is likely, or other contexts that are not always well defined. This is particularly relevant to the material from the Bala Hisar of Charsadda and the Northern Valley sites, where the material recovered is from pits, ditch fills and other contexts that may be considered less than ideal for extensive analysis.

In her discussion of the quantitative analysis of plant macroremain assemblages, Pearsall (1989, 195-6) gives three provisos: "(1) do not use any statistical technique that you do not fully understand (2) begin with simple tabulations and then apply more complex techniques, and (3) do not use approaches that require more rigor than the data are capable of sustaining". The last two points are particularly important in view of the quantity and type of data being analysed in this study. Here we are comparing small amounts of botanical and zoological data from both recent and older excavations and sampling strategies, where they are known, are less than ideal.

Therefore, the new botanical material has been identified and recorded, but the size of the assemblage means that further analysis beyond recording presence would be meaningless. The plant data that are published are also small in number, and consequently analysis has been limited to noting presence. When the material from the published sites and the Bala Hisar are considered in terms of phases and archaeological context, it is clear that there is not enough material from either the phases or the contexts to allow any further meaningful quantification. Ghaligai and the Bala Hisar are the only sites with material from more than a single period recorded, and the quantities from both are very small (see Appendices 2 and 4). Presence analysis has been used as the starting point in many environmental studies from South Asia, including an examination by Thomas et al. (1997, 770) of the animal remains from the Harappan sites of Kuntasi and Shikapur, Weber's (1992, 288) analysis of plant remains from the Harappan site of Rojdi, and also Belcher's (1991, 113) study of fish remains from Harappa.

The new zoological material was analysed by counting the Number of Identified Specimens (NISP) and estimating the Minimum Number of Individuals (MNI) recovered and identified, which allowed comparison with similar calculations from the published or in press information from the Swat and Dir Valley sites. The Number of Identified Specimens (NISP) is the number of identified bones and bone fragments from any given sample, be it context, phase or site (Davis 1987, 35-6). NISP is useful when comparing material from modern excavations with material from older sites (Bond 1994, 270) as it allows a direct comparison regardless of other analyses, which may not be directly comparable. Minimum Number of Individuals (MNI) is a calculation of the minimum number of individual animals for each identified type that a bone assemblage represents, and is useful in determining the relative numbers of types in an assemblage, and quantifying change in assemblage composition over time (Davis 1987, 36; Hesse & Wapnish 1985, 113). Both MNI estimates and NISP counts were used by Belcher (1991, 112-3) to assess the importance of fish and fishing at the sites of Harappa and Balakot.

2.5 The ethnographic material

The purpose of ethnographic research and its value to archaeology will be examined in the following section, and some methodological problems encountered in applying ethnographic observation will noted. The main methodological approaches used in this research will be presented, including sampling, interviews, analysis and dealing with uniformitarian assumptions in application to the archaeological material and questions. Ethnography is defined as the descriptive study of ethnic groups utilizing participant observation, or other form of fieldwork (Barnard & Spencer 1996, 193). Ethnoarchaeology is based on the premise that certain types of behaviour leave tangible signs in the material record, and that modern activities in the same area resulting in material remains can be used to aid interpretation of the archaeological record (Kramer 1979, 1). This is of course dependent on similarities in the environment, technology, and other factors between the past and the observable present, and so requires an assumption of uniformitarianism.

Allchin (1994, 1) has discussed the importance of recording what she calls 'traditions' relating to both social and economic practices throughout the whole of South Asia before these are lost through increased mechanisation and development. However, there also needs to be a distinction between ethnographic research, and anecdotal description. Ethnographic research is largely qualitative, and is based on quantitative research in the form of surveys and experiments (Hammersley 1992, 11). Work by Parkes (1987) in the Kalasha valleys in Chitral District of NWFP is an example of ethnographic research based on extensive observation and interview. Rahman's (1969) discussion of Swati customs and behaviour is an example of anecdotal ethnological interest. The work of Barth (1956), studying the different ethnic groups in the Swat Valley draws heavily on the ideas of animal ecology, in particular determining which particular ecological niche different ethnic groups occupy. Allchin points out that the use of modern analogy as a means of understanding the remote past has itself a history in South Asia (1994, 5), noting that, among others, James Ferguson and Alexander Cunningham in the 1870s, and Marshall in the 1930s were making use of contemporary traditional activities to interpret historic and prehistoric artefacts.

Ethnographic information is included in this research for three main reasons. First is the recording of such information for its own intrinsic value. Following from Allchin's (1994, 6) discussion of the importance of both recording and applying the examples of 'traditional' techniques and lifestyles, it is considered that interviewing a range of pastoralists and agriculturalists within the Charsadda District and the Valleys of Swat and Dir is a valid aim. Change is continuous, and as will be discussed below, many factors have and are affecting the practice of both pastoralism and agriculture in this region. Therefore, recording at this time can be very useful both to compare with past observations, and to use in future work. There are still many farmers practising 'traditional' or largely non-mechanised agriculture within the research area, and many pastoralists following 'traditional' or pre-motorised seasonal routes with flocks. According to the 1990 Census of Agriculture in NWFP, only 20% of farmers working less than 5 acres in the Charsadda District reported using fertilizer, whereas those using both fertilizer and manure was over 70% (Government of Pakistan 1994, 530). In Dir, only 2.5% of farmers on similar sized farms reported using fertilizer (ibid., 535). Tractors in Swat were used by 12% of farmers, contrasting with 72% who used draught animals (ibid., 710). These figures suggest that modern farming aids such as tractors and artificial fertiliser are not as important as the use of traditional approaches.

Second, the use of ethnographic data acquired through interview and observation, will inform the interpretation of the new and unpublished archaeological environmental material. The importance of ethnographic material in archaeological interpretation should not be underestimated: "If any statement about the past is unavoidably made in the present, it is also unavoidably *an analogy*" (Johnson 1999, 48). Accordingly, interpretation of past material remains and behaviour is based largely on our understanding and experience in the present (Hodder 1982, 11). The work of Mughal (1994), was primarily aimed at understanding the cultural sequences and development in the Cholistan Desert of Central Pakistan, during the Hakra and Harappan periods. In addition to this, however, he also studied the modern nomadic and sedentary groups in the region, and this information was used to support the development of his archaeological model of contact between the two economic groups (ibid., 53, 56, 64). In this volume, given the small quantities of archaeological environmental data the inclusion of ethnographic information is considered very important.

As a means of bridging the gap between observable, empirical data comprising the archaeological record, and developing theories explaining and interpreting these data in terms of past human behaviour, Binford (1989, 270) emphasises the need to develop middle range research. Middle range theory or research aims to make explicit the assumptions that researchers use for their interpretations of the past, and also to formalize these assumptions into testable theory (Binford 1989, 20; Johnson 1999, 50). Observing and interpreting certain activities and their products in the present offers one way of developing an explicit middle range theory. In his study of a group using stone tools in the central Australian desert, Binford makes this link between past and present clear, where the "aim was to study the relation between statics and dynamics in a modern setting. If understood in great detail, it would give us a kind of Rosetta Stone: a way of 'translating' the static, material stone tools found on an archaeological site into the vibrant life of a group of people who in fact left them there" (1983, 24). Further, the use of ethnographic observation and analogy must also be subject to repeatable testing procedures thus becoming scientific in approach (Hodder 1982, 14).

Third, ethnographic studies have been used to generate predictive models in terms of plant and animal remains. In environmental archaeology, some of the better known examples of these type of models are those of Hillman (1981, 1984) and Jones (1984), detailing crop processing by-products from Turkey and Greece respectively. In these studies, the very specific remains obtained from different processing stages have been used in the development of consumer-producer site predictive models. While at this stage in this research, such a study of a single element of production has not been possible, these examples show the potential to use modern studies in interpreting very specific elements in the archaeological record, and to draw conclusions about the nature of economic activity at a site. Reddy (1991, 1997) has developed similar models of crop products and by-products at each processing stage for the millet crops which are frequently encountered in South Asian archaeobotany, and there is an ongoing debate in archaeobotany about the applicability and validity of such models, particularly in different geographical areas. Nevertheless, they have been seminal studies in the use of ethnographic observation, interview and experiment as a means of interpreting archaeological material. That is, by recording products and by-products from such processing, it is possible to build a model of what is likely to survive, and thus be recovered, and in turn, what such remains could suggest about production, processing, trade and consumption. Other examples of the use of ethnographic techniques to develop models for the interpretation of environmental remains include the work of Halstead (1990), which has examined modern transhumant patterns in the Pindhos mountains of Greece, and applied his conclusions to archaeological interpretation. Palmer (1998) has studied extensively the effect of traditional farming practices on the weed assemblages associated with crops in Jordan, and from this and similar work, a greater understanding of traditional agrarian practice has arisen.

The ethnographic material collected during field research for this study is based on qualitative interviews and observation rather than the specific analysis of biological materials and other commodities given in the examples discussed above, and will be used to develop a general predictive model of subsistence for the area comprising the Northern Valleys of Swat and Dir, and the Charsadda District within the Vale of Peshawar, and this model will then be tested using the archaeological environmental data from the selected sites in these areas, and this will be presented in Chapter 9.

2.5.1 Collecting the ethnographic material

Interviews were carried out during two field seasons in Pakistan. The first of these, from April-May 1998 comprised interviews with settled farmers, whose primary support for their family and selves came from field cultivation, and non-sedentary groups with animals wintering around Charsadda. Further work in Dir and Swat allowed interviews with groups of pastoralists moving back up to their 'permanent' homes in the Northern Valleys, and some who had already reached their summer homes. During the second season, July-August 1999, interviews were carried out along the length of the Swat valley, and in the lower Dir valley. Figure 3 shows the location of the ethnographic interviews (page 212). The seasonal nature of the farming and movement in this area means that limiting the interviews to spring and summer has resulted in an unavoidable bias in the data. While this can be partially balanced by use of literature sources, it is also an area that could be addressed in future field work.

Other sources of bias include the location of the informants. Those interviewed at Charsadda were selected, among other reasons, because of their proximity to the archaeological site of the Bala Hisar of Charsadda. Those traveling back up from the Vale of Peshawar to Dir and Swat, were necessarily selected because of their presence on, or very close to, the road itself. Many informants said that the modern roads follow older routes between the hills and plains anyway, so this bias may be less significant at this stage than when locating informants in the valleys. In the Northern Valleys, as discussed in Chapter 6, the seasonal nature of the herding meant that many groups of potential interest had moved beyond reasonable access, and constraints of time and personal safety were other major considerations and limitations.

All interviews were carried out through an interpreter. Even when the interviewees spoke English, Urdu was used to clarify some details, and allowed the interpreter to both explain the project, and discuss the interviewer with the informant with some degree of privacy. The use of an interpreter has been viewed as both a positive and a negative aid to ethnographic fieldwork, and as interpreters are often, as described by Ellen (1984, 111), 'self-selected' or imposed from within the group being studied, their interaction with the group may greatly affect the reception and work of the interviewer. This is also true of interpreters who come from outside the community, as in the case of this work. Interpreters here were necessary not only for overcoming the language

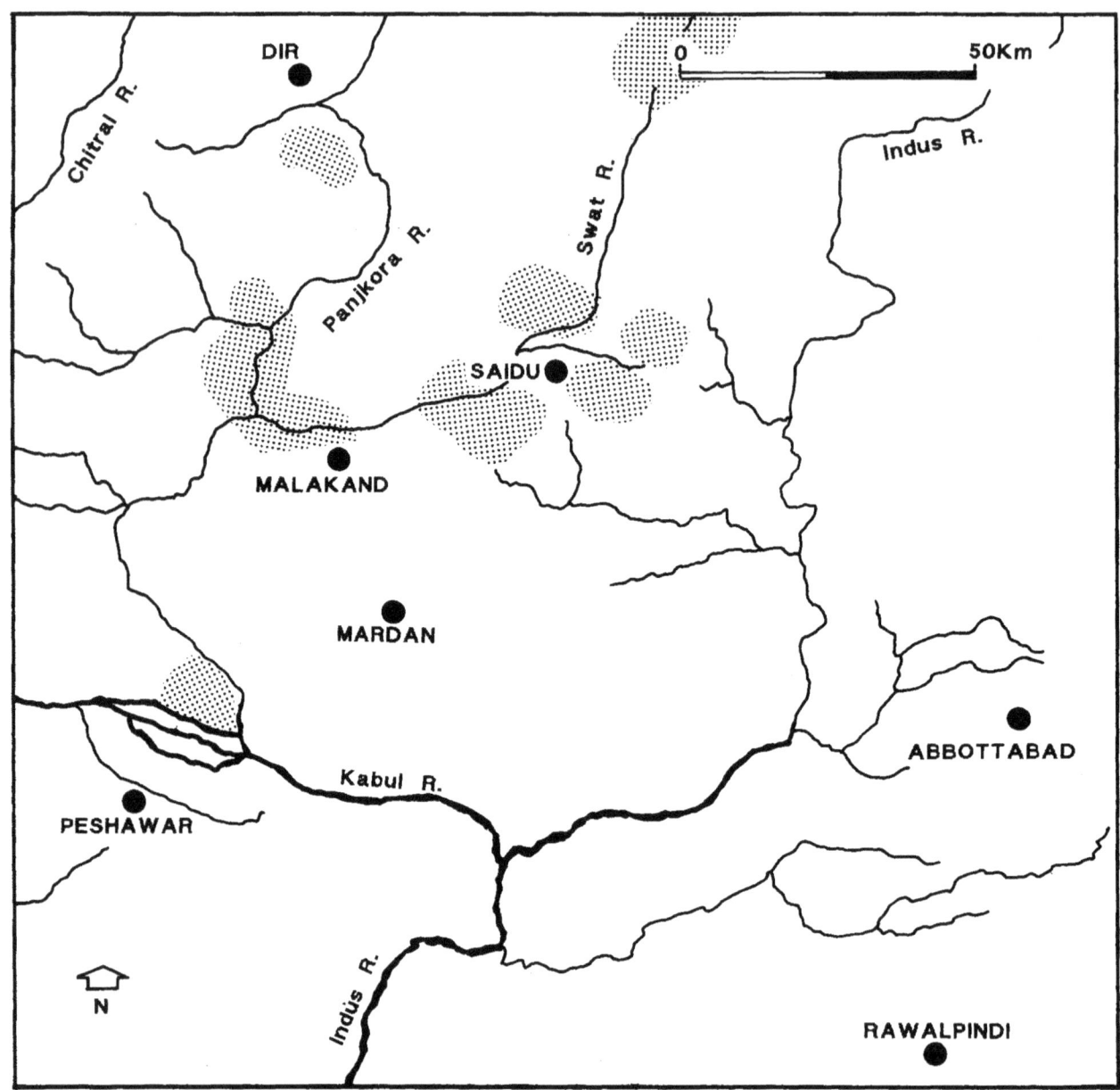

Figure 3. Location of ethnographic interviews (Alan Braby 2000)

barrier, but also to provide authority and protection in the field, and to locate informants. The interpreters for this study included the Professor of Archaeology at the University of Peshawar, and lecturers from the Department of Archaeology, all involved because they had some degree of interest or connection with the regions studied. While working in Dir and Swat, local residents known to the interpreters were also co-opted to the team to help locate suitable informants.

Just as the interviewer is acknowledged to bring to the field many pre-conceived beliefs and expectations (in both archaeology and ethnography), so too is the interpreter (Ellen 1984, 115-116). This extends to the subject of gender, which is important when a female interviewer is entering a Muslim, male dominated area, asking questions about herding and agriculture which, being conducted outside the house and compound, belong to the male sphere of activity. The farmers interviewed were Pathans, and so largely traditional Muslims practicing purdah. The accompanying interpreter (and other members of the retinue, varying in role and number) were also male. The prevailing feeling among many women conducting ethnographic fieldwork is that being female can confer a positive advantage, as in addition to being largely non-threatening "a female ethnographer can move between public and private domains more easily, and she may even gain access to men's sectors which are barred to women in the host society" (ibid., 124). This was certainly true in general in NWFP. While it was not possible to talk to women about their role in farming households because the presence of a male interpreter meant that we could not be admitted together into the company of women, the male farmers and herders, with one notable exception, were very willing to talk about their work. While I, as a woman, was made very welcome in purdah quarters, my lack of Urdu and the women's almost entire lack of English meant that we could not communicate in any detail. It was however, possible to talk to women in the mobile groups, as their Gujar ethnic affiliation allowed them to talk to the interpreters. In Chil Valley, Madyan, Upper Swat, men belonging to an extended family group were very reluctant to talk about their farming and herding. It was thought by the interpreter and the local guide, that this was because the farmers in question had formerly been mobile pastoralists, but had 'moved up' socially, and now identified themselves as part of a different ethnic group, and this is discussed further in Chapter 8 and Appendix 5, Interview 5.4.

All the informants claimed to have long family connections with their means of subsistence. Even when informants stated that while their fathers or grandfathers may have travelled between valley and plains, they remained in a single place all year, but were familiar with the past routines and activities. Allchin (1994, 1) points out the importance of establishing how old the traditional practices being studied are. Unfortunately, she does not proceed to outline a method for achieving this. All the farmers interviewed used little, if any mechanical farm equipment, chemical fertiliser or weed control. While the groups moving between valleys and plain owned, or hired trucks to move larger animals such as buffalo, and to transport elderly family members, they also had little if any machinery dedicated to farming.

The ethnographic interviews followed the same format, to ensure that the same questions were asked. These questions covered the following areas for sedentary farmers: family and ethnic grouping, length of time spent working or living in area, extent of land, crops grown and seasons, tools and techniques used for land preparation, crop management, harvesting and storage, animals kept on farm and their uses, sources of fodder, access to markets, and contact with mobile pastoral groups. The pastoral groups were asked the following: family and ethnic grouping, seasonal movement patterns, animals kept and purposes, animal husbandry techniques, crops grown and where and when, uses of crops, markets, storage, residences, and contact with sedentary groups. Two problems arose with the interviews. The first was that in both groups almost all the informants were working while the interview was taking place, and in particular, trying to ask detailed questions of someone who is continually darting off to rescue wandering goats, is rather difficult. Second, although the informants were almost all willing to talk to us, they could not really see the point of many of the questions, especially those concerning some of the details of storage or crop processing for example.

The purpose of these questions was to establish if there was a real difference between subsistence patterns and social organisation in the valleys and the plains. If so, how could these differences be characterised in terms of the plants and animals involved, and how could we then build both descriptive and predictive models from this information? Further, in terms of contact and reciprocity, it was intended to determine whether there were any patterns, and if so, how these might be distinguished in terms of material remains. If, as a result of analysing and comparing the subsistence strategies from each region, differences were distinguished, an assessment of the factors that might cause these differences would also be important to an understanding of the patterns.

One of the basic principles of ethnographic study is to move between descriptions, or the particular, to an understanding and development of the theory, or the universal phenomena (Hammersley 1992, 12-13). Therefore, it was the express aim of the work in this volume, both ethnographic and archaeological, to take individual descriptions of sites, site material and modern subsistence patterns, and develop a new theory, or interpretation of phenomena.

2.5.2 The ethnographic material and continuity

As noted above, ethnographic information is reliant on observations of 'traditional' farming, and the fundamental assumption of uniformitarianism in technique and method. This assumption is discussed at length in the literature in relation to the work of Hillman (1981) and Jones (1984) on crop processing (see e.g. Fuller 1999; Thompson 1996; van der Veen 1992). Without this uniformitarian assumption, there is little point in applying models based on modern data to archaeological data.

However, within this study area, a number of major changes affecting the ideological, political and physical parameters are known to have occurred, and there are undoubtedly many more between the early 2nd millennium BC and the late 2nd millennium AD that are unknown. While the length and aims of this particular work preclude an in-depth discussion of all these changes and their likely effect on the subsistence strategies of the inhabitants of the Vale of Peshawar, Dir and Swat, it is important to be aware of some of the major events, and how they have may affected economic organisation. These points will be dealt with in greater depth in the next chapter, as part of the description of the physical and social setting of the study. Political events such as changes in leadership, state formation and the change in status of provinces or independent states, Government control of land and issues of land ownership, irrigation provision and religious and ideological change could have had a potentially significant effect on subsistence strategies.

One method of assessing the possible impact of such events, particularly those in the more recent past, is by comparing modern information with historic records. For example, the ethnographic material can be compared with the information from historical sources such as the district Gazetteers. A recent Census of Agriculture report (Government of Pakistan 1994) for all the areas in the study is also a source of information about modern land use and practices, and again these data can be used as a comparison with the ethnographic material and the historical sources. Gazetteer material from the turn of the century is available for the Peshawar District, and this includes Charsadda District. As there are no comparable Gazetteers published for Swat and Dir, relevant information from other historical accounts such as Stein (1929) and McMahon and Ramsay (1901) will be used.

Recognising the possibility for discrepancy between historical and archaeological records is important. Schopen (1997, 8, 14) investigates this issue at some length in relation to Buddhist doctrine and practice, and highlights not only the differences that exist, but also some of the possible reasons for this situation. In an attempt to develop a methodology for exploring ideological change through material remains, the archaeozoological and the archaeobotanical material from the site of Anuradhapura in Sri Lanka were analysed to determine whether they reflected the historically attested change from Hindu to Buddhist practice on the island (Coningham & Young 1999). A parallel investigation of craft working was also carried out, but neither were able to clearly show any major changes that could be linked to this fundamental change in ideology and social organisation.

In NWFP a number of major physical changes have occurred, including the building of the Swat Canal and extensive deforestation, and the effects such changes have had on pastoralism and agriculture also need to be examined. Allchin (1985, 27-28) notes that both the Badwal nomads of central India and the nomadic groups of NWFP and Baluchistan are facing major changes that have the potential to affect their seasonal movements. These include the building of railways, bridges, settlements, tectonic activity that could alter the course of rivers, grazing and so forth, political changes and frontier closures. However, despite these often major events, Allchin asserts that "in both cases it is the general pattern of movement and the habit of mobility that are significant" (ibid., 28). Again, there is no move towards quantifying the effect of such change on subsistence patterns, or developing a methodology to do so. It is unfortunate that so many studies utilising 'ethnographic studies' in both South Asia and in other areas of the world, fail to recognise the importance of addressing discontinuities in the past of any group that may have an effect on the behaviour being observed in the present. For example, Maclean and Insoll (1999), in their study of archaeological, historical and ethnographic food and 'cuisine' in Mali state explicitly that they are aware that "fundamental elements of cultural identity are changing at an increasing rate" (ibid., 79). Yet nowhere in this study do they attempt to explore and quantify these changes, and assess the real effect they have had on the traditional 'cuisine' of Gao and Mali.

2.6 Summary and conclusions

There are many acknowledged problems in dealing with small faunal and botanical assemblages (Meadow 1991; Reid & Young 2000), and these have been briefly discussed here in terms of both the restrictions and possibilities of interpretation which small assemblages present. The method of data collection at all the sites contributing to this study and the small assemblage size here intensifies the problems that are faced in the analysis of any environmental data. These problems include taphonomic questions, preservation issues and sampling strategies. It also means that distinguishing trends within each site, and making comparisons between sites is difficult.

However, the presence of both faunal and botanical remains means that it is possible to make statements about the presence and absence of species, and their relative quantities. This information about resource exploitation can then be used to make comparisons between the different types of sites. Utilising information about the ecological conditions of both plants and animals can help in understanding more about the environment of the site, and perhaps about conditions and constraints affecting subsistence strategies.

Integration of two sources of environmental data to develop a more comprehensive picture of the subsistence base and resource exploitation of the Late Bronze and Iron Age inhabitants of rural and urban NWFP is an important step forward in the archaeological interpretation of this region. Within modern environmental archaeology this sort of data integration within an interpretative framework is seen as a very useful means of understanding more about the economic and social organisation of a site (e.g. Reddy 1991, 1997; Thomas 1999). To undertake this type of analysis with data from the sites of Aligrama, the Bala Hisar of Charsadda, Balambat, Bir-kot-ghundai, Ghaligai, Kalako-deray, Loebanr III and Timargarha, is not only a necessity due to the quantity of available data, it is a means of obtaining maximum information at this time from the assemblages. By bringing in a third source of information - ethnographic studies related to subsistence and land use - the understanding of the archaeological data through the ability to develop predictive and interpretative models is increased greatly.

It is important to record and interpret the data that are available, and this work shows how environmental material can be used to further understand economic and cultural issues within the aim of gaining a greater understanding of the subsistence strategies of the two study regions. Such a collection of data can be added to when new sites are excavated and sampling for environmental material is incorporated into the excavation strategy, and its subsequent analysis and interpretation is incorporated into the overall research design.

Chapter 3
Physical environment and social identity

There are two sections in this chapter; firstly a description of the physical setting of this study and characterisation of the two regions involved, and secondly, a discussion of the social identities of the inhabitants of both. These are important not only to understand the physical contrast between the Northern Valleys and the Vale of Peshawar, but also to determine the effect this has had on subsistence strategies in the different regions, and the influence of such factors as ethnic group. As discussed in Chapter 1, the interpretation of the archaeology of these two regions has been based on a largely artificial separation between them, and has resulted in their treatment as culturally distinct, with little or no contact and exchange between them. Following this, there is a section addressing the major areas of change with respect to continuity and questions of the applicability of uniformitarian assumptions in applying ethnographic study to archaeological material, and a comparison of modern and historical data relating to land use will allow an assessment of the significance of some of the recent changes.

3.1 The Vale of Peshawar

The Vale of Peshawar or the Peshawar Basin is a flat alluvial plain with hill ranges on three sides, and the Indus River to the east on the fourth side (Punjab Government 1897-8, 13). This basin is in part the result of the Kabul River and its tributaries running across the plain, and covers an area of 6,500 square kilometres in the shape of an ellipse, which runs 120 kilometres in width from east to west, and 86 kilometres in length from north to south (Dichter 1967 91-2; Imperial Gazetteer of India 1904, 144). The Vale of Peshawar is open only to the east where the Kabul and Indus Rivers meet, and from here the land then slopes steadily up to the Potwar Plateau in the Punjab (Dichter 1967, 91). Access on the south west and north is via passes through the hill ranges, such as the Kohat Pass leading south to Khattack and southern NWFP, the Khyber Pass leading through the Khyber hills to Afghanistan, and north over the Malakand Pass to Chakdara and from there to Dir and Swat.

Charsadda town and the archaeological site of the Bala Hisar of Charsadda are located in the flood plain created by the Swat River joining the Kabul, and this has important implications for both flooding and soil fertility in this region. The Swat River, being mainly fed by glacier and snow melt, has a very powerful flow in the summer, but is much less in mid-winter. This is important in terms of water availability for crops, where irrigation is needed in spring and summer, for the autumn harvested crops, and when rainfall decreases in winter, more irrigation is required for the crops sown in late autumn for harvesting in the spring (Imperial Gazetteer of India 1904, 116). These cropping regimes and irrigation are discussed in greater detail in section 3.4 Land use, below.

The presence of the Kabul River and its tributaries such as the Swat, has resulted in the alluvial soils that combined with irrigation has given rise to the intensive agriculture practised here today. These soils are part of the Indo-Gangetic spread of alluvial soils and "Economically, they are the most important soils in South Asia, providing the basis for settled agriculture for millennia" (Robinson 1989, 22). Alluvial soils are however, sometimes subject to waterlogging and salinity problems, and waterlogging of irrigated farmland has been reported in the village of Gara Tajik north of Peshawar City, as a result of the Warsak Dam (Robinson 1989, 23; Shaukat & Begum 1992, 2). The Charsadda Tahsil or District itself was described as an area of great fertility and highly cultivated, with more woodland cover than many other parts of the Peshawar Basin, and its low lying position by the Kabul and Swat Rivers resulted in plentiful irrigation (Punjab Government 1897-8, 19).

The Vale of Peshawar is characterised as semi-arid, with no distinct season of maximum rainfall. Its continental position, away from the modifying influence of the sea, along with low humidity levels, means that it is subject to extremes in temperature. The mean annual rainfall in Peshawar City itself is between 250 to 370 millimetres, and across the Vale is between 250 and 500 millimetres (Dichter 1967, 12-13, 22). The Vale of Peshawar has four main climatic periods or seasons. The cold weather period is from December to March, when there is some frost and daytime temperatures average 15°C. The months from April to June are described as a transition period, characterised by a rapid rise in daytime temperatures over the months, and from the middle of May, daytime temperatures average 40°C. July to September is the monsoon, and while there is a slight temperature drop, this is accompanied by an increase in humidity. The final climatic period is another transition phase, spanning the end of the monsoon in October through November to the beginning of the cold weather again. During this transition period there is little rain, and the temperatures are decreasing (ibid., 104-6; Imperial Gazetteer of India 1904, 83).

Lowland NWFP has been characterised as dry and sub-tropical in terms of the vegetation and fauna (Roberts 1997, 13). The dominant tree cover in the Vale of Peshawar includes a mix of acacia (*Acacia* spp.) and olive (*Olea* spp.), types that are suited to arid conditions (Edlin et al. 1978, 202; Mabberley 1987, 2, 408), and there is also grass and scrub cover. A Gazetteer (Imperial Gazetteer of India 1904, 145) records the presence of mulberry (*Morus* spp.), shisham (*Dalbergia sissoo*), willow (*Salix* spp.) and tamarisk (*Tamarix* spp.) trees in irrigated areas, and acacia and scrub in dry areas (Mabberley 1987, 383, 537). It was also recorded in 1901, when NWFP was formed, that "there is ample historical evidence that in ancient times the District was far better wooded than it is now, and the early Chinese pilgrims often refer to the luxuriant growth of trees on hill-slopes now practically bare" (Imperial Gazetteer of India 1904, 152-3). Chinese pilgrims such as Fa-hsien recorded his travels to this region during the 5th century AD (Stein 1929, 13).

There are virtually no large wild mammals found in the more densely populated and cultivated areas of the Vale of Peshawar today, which includes the Charsadda District. The main exception to this is wild boar (*Sus scrofa* L.), as the cultivated sugar cane provides them with good cover (Roberts 1997, 235). Wild pigs are described as a pest of cultivated land and are present in forest up to around 900 metres. Other smaller mammals include the porcupine (*Hystrix indica* L.), also a known pest of cultivated land, and very adaptable in terms of its environmental requirements. The porcupine is reported as being hunted for food in parts of Sindh (ibid., 339).

3.2 The Northern Valleys

The Swat Valley extends along the Swat River from just north of Chakdara to Kalam. Above Madyan however, the valley contracts and the river flood plain is lost. Therefore, the valley can be divided into two main areas: Lower Swat, up to Madyan with the river flood plain, and Upper Swat including Swat Kohistan above Madyan. The Swat River rises at Kalam at a height of 2,000 metres and when it reaches Madyan it has dropped to 1,300 metres, which is a drop of approximately 30 metres per kilometre, indicating how steep the valley is at this point. Between Madyan and Landakali to the south, the valley is on average 3-5 kilometres wide with the Swat River extensively braided, and it is in this lower part of Swat that settlement is greatest (Dichter 1967, 32, 52; Imperial Gazetteer of India 1904, 217). Flooding of the river deposits silt on the surrounding land, adding to its fertility, and seasonal khwars, or tributary streams, run down from the hills into the main river. The minor ranges of the Hindu Kush extend south into both Dir and Swat, with heights of around 5,500 metres in the north and 1,500-1,800 metres in the Malakand area (Dichter 1967, 8-9). The highest peak in Swat is thought to be over 5,800 metres, indicating the potential ruggedness of the terrain (Mian 1956, 61). The main access routes in and out of Swat are through a series of passes over the ranges. To the south, the Malakand Pass leads to the Vale of Peshawar, Shangla Pass in the east leads to the Indus Valley and river, Badwai Pass in the north west leads to Dir Kohistan, Bisao pass in the north east leads to Indus Kohistan and Kach Koni Pass in the north leads to Chitral (Dichter 1967, 49).

Like Swat, the Dir Valley follows the course of the main river, here the Panchkora. The Panchkora rises in the Shandur Range of the Hindu Raj, which extends south west from the Hindu Kush to the Lowari Pass (Dichter 1967, 27). The river again is the main focus of settlement in the valley on the alluvial flood plains, and there is a similar very steep rise in altitude in the valley, from between 1,500-1,800 metres in Lower Dir to over 3,000 metres in Upper Dir. However, Dir has been described as much more rugged than Swat, with many more narrow gorges through which the river passes, and these also occur further downstream than along the Swat river, and this necessarily limits the amount of cultivable land in the valley. Agriculturally, the alluvial river plains are the most important areas in both valleys, and the majority of settlement is concentrated around either river or the larger tributaries. However in Dir, the Panchkora River is thought to carry a much heavier load of silt, and this has an affect on both erosion and soil fertility (Dichter 1967, 7). Agriculture in Dir has also been very badly affected by deforestation and related erosion (ibid., 34). Passes over the hills allowing access in and out of Dir include the Malakand to the south, however north of Chakdara there is another pass, the Katgala or Cut Throat Pass, which divides the Swat and Panchkora watersheds (ibid., 63). To the north the main route into Chitral is the Lowari Pass, but there is also the Tal Pass in the very north of Dir, and the Badwai and Zhandrai Passes lead east into Swat Kohistan (ibid., 67). While the Malakand Pass which gives access to both Swat and Dir from the Vale of Peshawar is generally passable throughout the whole year (c. 1500 metres), passes further north are seasonal (Spate & Learmonth 1954, 490). Indeed, the Lowari Pass (3,118 metres) which gives access from Dir to Chitral is closed between October and May, due to both snow and flooding (Haserodt 1996, 9).

In terms of climate, Lower Swat is cooler than the plains in summer, but has higher humidity as the maximum precipitation is generally recorded in the summer. Lower Swat is semi-arid, the lowest temperatures between 0°C and 10°C, and its highest between 20°C and 30°C. Upper Swat is also semi-arid with maximum precipitation in summer, but with a lower temperature range, with its lowest below 0°C, and its highest temperatures between 0°C and 10°C (Dichter 1967, 12, 22). Annual rainfall also varies from an average of between 750-1000 millimetres in Lower and Mid Swat, to between 500-750 millimetres in Upper Swat. As the Swat River is subject to the influence of both monsoon rainfall and snow melt, flooding is a serious problem in the valley (ibid., 32). Snow on the valley bottoms in Lower Swat is relatively rare, but winters in Upper Swat and Swat Kohistan can be very severe, with snow present for more than nine months of the year (Mian 1956, 61).

The climate of Dir is similar to that of Swat, but with lower humidity levels. Dir can also be divided into two main climate zones, with Lower Dir described as semi-arid with summer maximum precipitation and lowest mean temperatures falling between 0°C and 10°C and the highest between 20°C and 30°C. In Upper Dir the mean lowest temperature is below 0°C, and the mean highest between 10°C and 20°C (Dichter 1967, 22). Average rainfall in Lower Dir falls between 750-1000 millimetres and in Upper Dir between 500-750millimetres (ibid., 12). Dir Kohistan suffers from severe winter conditions with extended snow cover, while in Lower Dir, snow is known to fall in the valley bottoms, but this is not frequent (Mian 1956, 60).

Swat and Dir can both be divided into a number of zones relating to vegetation and animals. In the north, the vegetation of Swat Kohistan is composed of alpine meadow and sub-alpine scrub with birch forest. Swat also has areas of dry, temperate coniferous forest, notably around (Malam) Jabba, and the other high places in mid-Swat. These forests are dominated by spruces (*Picea* spp.), the Bhutan pine, *(Pinus wallichiana)*, which is native to the region stretching from Afghanistan across to Nepal, and the deodar (*Decrus deodara*) (Edlin et al. 1978, 125; Mabberley 1987, 109, 452). The deodar tree has been of economic significance in both Dir and Swat for many years (Imperial Gazetteer of India 1904, 216-7). Lower Swat, reaching as far south as Malakand has 'steppic' forest cover comprising juniper, pine, and scrub and grass cover of sea wormwood (*Artemisia maritima* L.) and *Ephedra nebrodensis* (Mabberley 1987, 207).

The Dir Valley falls into a broad vegetation zone of sub-tropical pine forest and dry sub-tropical and temperate semi-evergreen scrub forest (Roberts 1997, 12). This can then be divided into vegetation categories, that, like Swat, start with Dir Kohistan in the north which is a region of alpine meadow, sub-alpine scrub and birch forest. Mid Dir is an area of dry, temperate coniferous forest, characterised by juniper and pine, with artemesia and ephedra grass species. Evergreen oak and pistachio have also been noted here (Roberts 1997, 12), and Lower Dir has steppic forest cover.

The wild mammals still found in Dir and Swat include a number that are adapted to severe winter conditions, including the red bear (*Ursus arctos* L.) and the high mountain vole (*Alticola roylei* Gray) (ibid., 153, 431). The Himalayan grey goral (*Naemorhedus goral* Hardwicke) (up to 6,500 metres) and the markhor (600-3,600 metres) are also found at different altitudes, but similar ecological niches (ibid., 265, 283). For these two animals, the differentiation by altitude is an important means of ensuring they are not competing. The migratory hamster (*Cricetulus migratorius* Pallas) is found in both Dir and Swat in the steppic cover regions, and is known in areas of terraced cultivation, but is not known to damage crops (ibid., 403-4).

3.3 The people

While the Vale of Peshawar is the most densely populated area, with a population density in 1972 of between 2,590 and 3,800 people per square kilometre (Kureshy 1977, 75), Swat is the most densely settled of the political agency areas of NWFP. In the 1951 census a population of 518,598 and an area of some 6,500 square kilometres gave a density of over 80 persons per square kilometre, reported as more than twice the average density for all the other tribal regions (Mian 1956, 61). This figure had risen to over 1 million in 1982, thus doubling the density, and while figures were not available for Dir, the population there was reported as much lower and much less dense (Lindholm 1982, 23). This problem with figures from Dir is highlighted by Mian (1956, 60) who reports that the 1951 census figures for Dir give a population of 148,648, however in a footnote he explains that "it is claimed by the State that its actual population is not less than 500,00 people". Using the figure of 148,648 for population and that of 4,800 square kilometres for area, the population density for Dir is then 31 persons per square kilometres, which is less than half that of Swat. It is not clear whether these figures include mobile groups, or whether only settled groups are counted. Mughal (1994, 53) mentions the difficulties of obtaining accurate figures for nomadic groups in the Cholistan Desert in east central Pakistan.

What is clear from any study of ethnic groups in NWFP is that there are a number of different groups, and "In so far as the term 'ethnic groups' has meaning in Pakistan it does not refer to large national minorities distinguished by a combination of race, language and culture. It refers, instead, to fairly small hereditary social groups defined in a variety of ways" (Wilber 1964, 53). A Gazetteer based on data gathered up to and shortly after the creation of the North West Frontier Province in 1901, also agrees with these assertions about the ethnic mix of the area where "the population contains several ethnological strata, representing the deposits formed by different streams of immigration or invasion" (Imperial Gazetteer of India 1904, 32). Within this mix, Pathans account for the majority in the whole population, and also for the majority of the agricultural population (ibid.).

Pathans or Pukhtuns (in particular the Yusufzai tribe) dominate the populations of both Dir and Swat (Lindholm 1982, 22). Indeed, Pathans are the majority throughout NWFP, and Pathans are one of the largest tribal groups in the world, spreading across southern Afghanistan (Weiss 1995, 109). The Vale of Peshawar is almost entirely Pathan today. Dichter's (1967, 95) brief account of the Salim Khan village situated off the main Mardan-Charsadda road gives the entire (500 or so) population as Pathan, and most are Sunni Muslim (Weiss 1995, 111), showing their dominance in terms of the ethnic identity of the farming groups around the Charsadda District. Ahmed (1991, 7) says that Pukhtun society can be broadly divided into two groups, the first of which is essentially egalitarian in nature, while the second is far more hierarchical. The first of the groups tends to be located in the low-production areas of NWFP. In contrast, the second group is found on irrigated, hence more productive lands, and due to the ranking within its composition, is usually part of a bigger state system. According to Ahmed (ibid., 7), nang or honour is one of the most important symbols of the egalitarian groups, and qalang, or the idea of taxes and rents, is significant for the ranked groups.

Among those who have studied the ethnic groups of NWFP there is some consensus and also some disagreement about these groups and their occupations. For example, Wilber (1964, 60) says that Pathans (or Pukhtuns) are in the majority, but also describes Sayyids, a specialised religious group, and Awans and Gujars who are "agriculturist" in nature, and nomadic Pashtuns or Powindehs, who move between Afghanistan and Pakistan with herds of sheep and goats. The 1897-8 Gazetteer also lists a number of different ethnic groups present in the Peshawar District. While Pathans dominate, and are described as now agricultural and settled, they are also thought to have previously been nomadic shepherds (Punjab Government 1897-8). In addition, there are groups of Hindkis, of Indian or Punjabi origin, and Gujars. The latter are here described as the original inhabitants of lowland NWFP (ibid., 125, 141-3).

Barth (1956, 1079) says there are three major ethnic groups within Swat state: "(1) *Pathans* – Pashto speaking (Iranian language family) sedentary agriculturalists; (2) *Kohistanis* – speakers of Dardic languages, practicing agriculture and transhumant herding; and (3) *Gujars* – Gujri-speaking (a lowland Indian dialect) nomadic herders". According to Barth, Gujars in Pathan dominated areas are the main practitioners of transhumance, moving between southern Swat and the northern mountains. Barth also says that the distance covered by transhumants varies locally, and those travelling in regular seasonal patterns with sheep and goats he designates nomads (ibid., 1083, 1085). This contrasts directly with Wilbur (1964), who says that Gujars practice agriculture, and indicates something of the confusion about ethnic groups and their subsistence practices in this region. Stein (1929, 96, 113, 131-2) variously describes the settled residences of Gujars, the summer shelters for Gujar families up on high mountain pastures where they bring their cattle for grazing, and Gujars who cultivate land owned by others. The summer mountain settlements also have provision for both grazing and cultivation (ibid., 115).

Dichter (1967, 54) notes that in Kohistan, there are two main groups. First the Gujars who are 'nomadic' landless people supporting themselves with herds of buffalo and cattle, and second the Ajars, also 'nomadic' who have herds of sheep and goats. Some Gujars are now thought to be settling, often as tenants of Pathans. Pathans are primarily settled agriculturalists and are the majority south of Kohistan. Dichter also describes the Kohistanis as practicing at least two types of transhumance. There is less discussion about the ethnic groups of Dir, but it is generally held that Dir is on the whole much poorer than Swat, with many landless poor being forced to leave (Dichter 1967, 66; Lindholm 1982, 23). This suggests that while the population of Swat is more dense than that of Dir, the quality of land allows more people to be supported. It may also be that in Dir there is a greater mobile, and thus uncounted, population. Indeed, the Sakas from Central Asia, and Indo-Parthian rulers of Gandhara during the last centuries BC, are thought to have themselves been nomadic pastoralist groups (Caroe 1958, 71; Imperial Gazetteer of India 1904, 14).

This is not intended as an exhaustive study of ethnic groups in the Vale of Peshawar and the Northern Valleys, but rather to characterise the main modern population trends within a very complex area. It is clear that there is no easy agreement about ethnic groups in the recent literature, and Chapter 8 will look in greater detail at modern observed groups and their subsistence practices.

3.4 Land use

Understanding exploitation of the land is central to this study, and an awareness of modern and historical cultivation can contribute to this. By examining the way land is used at present and has been used in the recent past, it should be possible to compare the two areas being studied and elucidate trends that will point towards similarities and differences. From this, it is possible to make statements about the important parameters that constrain subsistence in the different areas. Further, information relating to agriculture and animal husbandry in modern Charsadda District, Swat and Dir, and historical data and accounts, should give a relatively objective data set that can be used as a check on the consistency of the modern ethnographic information.

In 1903-4, the total area of the province was given as 34,400 square kilometres, and the total cultivated area was given as 11,000 square kilometres (Imperial Gazetteer of India 1904, 85). The importance of agriculture in NWFP is further shown by figures from the 1901 survey, where 64.5% of the population were recorded as dependent on agriculture (ibid., 34). When the diet of urban and rural people was analysed, the townsfolk staple was reported as mainly bread made from wheat, while rural diet changed according to the seasons. In summer, wheat and barley cakes were eaten, along with vegetables, herbs, wild fruit and milk products. In winter, the rural population consumed maize and millet based foods, which would have been harvested as autumn crops (ibid., 34-5; Punjab Government 1897-8, 101). Rural and 'lower' classes were reported as seldom eating meat, while 'better' classes ate meat, fowl and rice (Punjab Government 1897-8, 101).

The modern intensity of agriculture in the Charsadda District is indicated by the fact that between 91 and 98% of all farms have their total area under cultivation. In Swat this figure drops as low as 44%, and in Dir, some of the farms have as little as 13% of their total area under cultivation (Government of Pakistan 1994, 2, 7-8). Some of the reasons for this intensity of cultivation in the Charsadda District are likely to include pressure of population in the Vale of Peshawar, and the greater soil fertility here, resulting from both irrigation and use of fertilizer. In Charsadda District, 98% of the total area of land under cultivation is irrigated, which is in great contrast to the irrigated cultivated land in Dir and Swat. In Swat, only 50% of cultivated land is irrigated, and in Dir the figure is 80% (ibid., 159).

Irrigation in each area also varies according to the season and this affects the type of crop grown. In Charsadda, virtually all the summer (kharif), winter (rabi) and orchard crops are irrigated. In Dir, approximately three-quarters of summer, half of winter and almost all the orchard crops receive some form of irrigation. In Swat, the figures are far less. Here, around one third of summer, less than half the winter crops and around three-quarters of the orchard crops are irrigated. These figures can be related to the greater intensity of cultivation in Charsadda District, and also reflect the greater rainfall of Swat. Swat, surprisingly in view of its noted fertility and availability of suitable land, has the lowest percentage of wheat grown, with only half of this irrigated. Paddy cropping of rice is far more significant in Dir and Swat than Charsadda, and rice was noted as an export from Swat to the Vale of Peshawar, both at the end of the 19th century and early in the 20th century (Punjab Government 1897-8, 213; Stein 1929, 51-2). Fodder crops are much more important in Charsadda District. This reinforces the idea of limited grazing for animals within the densely cultivated area, and also that the type of animals kept, such as buffalo and cattle require fodder to produce better milk rather than simply grazing.

The historical data from the end of the 19th century also mention two crop seasons per year, with the kharif or summer crops sown between May and August, and harvested between September and December. The rabi or winter crops are sown between October and January, following the harvest of the kharif crops, and then harvested during April and May (Imperial Gazetteer of India 1904, 38). Sugar cane is called a winter crop, but in practice it is the ground for nearly the whole year as it matures. Wheat was the main winter crop in 1901, with barley the next most important, recorded as giving a slightly lower yield than wheat (ibid., 40). Maize was the main summer crop, with millet also a summer crop, but significant only in unirrigated areas such as D.I. Khan and Kohat (ibid., 40-1). Pulses were mainly mixed in with other crops, and cotton was grown around Peshawar, and its seeds were used for animal fodder (ibid., 41). Sugar cane was known, but accounted for a very small proportion of all the crops grown, in which cereal crops dominated (ibid., 85). The brief description of agricultural practice in the 1904 Gazetteer describes cattle pulling wooden ploughs fitted with iron blades, following which the majority of planting was done by broadcasting seeds. Little weeding was reported, and following harvest, the grain was trodden out by oxen and winnowed by hand with fans. Agricultural implements, with the exception of very modern ones for sugar processing, were described as of 'ancient type' (ibid., 39).

With regard to modern animal keeping, in Charsadda, cattle and buffalo account for 75% of the total domestic animals (Government of Pakistan 1994, 771). Domestic animals, for the purposes of the census are cattle, buffalo, sheep and goat. In Dir, female goats (over one year old) comprise easily the largest single category. Cows are the next most significant category, then bullocks. Buffalo are negligible and male goats and female sheep account for around 10% in each case. Modern Swat is more similar to Charsadda than Dir in terms of animal keeping. Cattle account for 43% of the total here, and buffalo for 23%, while goat account for 25% (ibid., 776-7). Reasons for this will be touched on in the ethnographic section, however, cattle types are not separated here, and it is not known whether the same types are present in the hills and on the plain. The flatter terrain and greater area of alluvial plain in Lower Swat are certainly more conducive to buffalo keeping than the rugged hills and gullies of Dir. In 1901, local cattle breeds in Peshawar District were noted as being small and weak. The exception to this was the Peshawar buffalo, which were described as remarkably strong and of great use for transport (Imperial Gazetteer of India 1904, 43). Sheep and goat were also present, and large numbers were recorded as coming into the region from across the border for winter grazing, but which border, and other details are not noted (ibid., 43). However, two different types of sheep were described, both the ordinary tailed, and the fat tailed or dumba variety (ibid.), both of which are found in the area today.

3.5 Factors that may have affected land use

Changes that may have had an effect on land use in this whole area include that of leadership and change to political organisation, religious and ideological change, and direct physical change such as irrigation development. However, the assessment of the effects of these changes here is reliant on recent historical sources, and in this area it is clear that major change in all of these areas has been occurring for many millennia, not least of which are the power struggles between Pathan tribes and rulers (Stein 1929, 3-5, 22-3). Proto-historical material by writers such as Herodotus and Arrian (Caroe 1958, 45-6) record major changes in rule and politics in this region from around the 6th century BC, the period of Achaemenid rule, and the subsequent invasion by Alexander the Great and his army in this region. Almost certainly the many changes in the control of this area are due at least in part to its position and significance in terms of access and movement of both goods and people. It has been suggested that "the lands which are now Afghanistan and the North-West Frontier of Pakistan have seen perhaps more invasions in the course of history than any other country in Asia, or indeed the world" (Caroe 1958, 25). While the changes noted in proto-historical and historical records cover many centuries, the archaeological record also shows change in prehistory in this region. For example, Dani (1967) notes that a clear change in burial patterns is evident at the site of Timargarha in Dir, and ascribes this to cultural change.

The British arrived in Peshawar in 1849, and the province of NWFP was created in 1901 by Lord Curzon, supposedly as a 'Pathan' province by merging the Districts of Peshawar,

Kohat, Bannu and Dera Ismail Khan, with the political agencies and tribal territories around them. The Malakand Agency was formed in 1895 (or 1896), and included the States of Chitral, Dir and Swat. In 1901, the control of this Agency was transferred from the Central Government of India to the NWFP administration (Caroe 1958, 383, 413; Imperial Gazetteer of India 1904, 215). Despite the good intentions of the British in India, Ahmed (1991, 29) says of their relations with the Pathans or Pukhtuns that "to the end it was a straightforward jihad. The retention of certain British traditions by the Pakistani administration must not be allowed to obfuscate the tribesman's rejection of British civilisation". Indeed, as Caroe reports, there was little attempt to discover or record what the Pathans themselves thought of the changes imposed by the British (1958, 414). However, on the subject of agricultural development and irrigation, Caroe says that from 1925 onwards, there was intensification of agriculture in Peshawar, Swat and Bannu districts, and over time "the ravages of the Sikhs were made good, and more than made good" (ibid., 429). This strongly suggests that while the 19th and 20th century AD developments in agriculture were very significant, they were not unprecedented in terms of agricultural output. Indeed, they may be viewed as part of an ongoing cycle of intensification and ravage, linked into political and social change.

In a study of the development of states in the Middle East, and the effect these had on tribal affiliations, Khoury and Kostiner (1991, 3) demonstrate that while state formation in the 19th century resulted in the breakdown of many aspects of traditional tribal life and affiliation, the new groups retained many characteristics and aspects of the former tribal groups. The idea that while there may be change, many fundamental beliefs and ideas stay the same is echoed by Caroe in his study of the origins and history of the Pathans when he says "There is plenty of evidence after we reach the period of documented history to show that the hill-tribes have been little incommoded by empires and the passage of armies" (1958, 42). Some of this continuity can be seen in place and tribal names recorded in Herodotus and comparable with those still in use (ibid., 41).

The advent of Islam is probably the most recent major ideological change to have occurred in the whole of South Asia. Like many religions, Islam has laws relating to food consumption and these might be expected to have changed attitudes to food procurement. Perhaps one of the best known dietary rules, that may be detected through change or absence in the archaeozoological record, is that forbidding the consumption of pig meat and products (Insoll 1999, 96). Using a case study of animal remains and the introduction of Islam in Jordan, and the association of specific plant foods with Islam, Insoll argues that it is possible to detect very subtle data trends to examine the archaeology of Islam through dietary remains (ibid., 97-100). However, a recent preliminary study of modern agricultural and pastoral activities within the Kalasha (non-Muslim) and Kho (Muslim) communities of Chitral, shows no fundamental differences in subsistence strategies or diet (Young et al. 2000, 138-139). Despite very different ideologies central to each group, it certainly appears that in this extreme geographical setting, environmental factors are at least as important, if not more so, in shaping economic decisions.

Prior to Islam, Buddhism is also recorded in both historical works and archaeology, to have been of great significance in NWFP. The accounts of Buddhist pilgrims such as Fa-hsien, travelling in the 5th century AD through to Hsuan-tsang in the 7th century AD, record Swat in particular as very fertile, and home to many Buddhist monasteries and monuments (Stein 1929, 13-4). According to Hsuan-tsang, prior to the White Hun rampages of the 7th and 8th centuries, there were in the order of 1,400 monasteries and 18,000 Buddhists, and Stein describes Swat as previously "thickly populated" in comparison with the time of writing (ibid.). Such accounts of population fluctuation suggest that while this area is fertile enough to support a relatively large, settled population, other factors and conditions almost certainly play a significant role in determining these levels, and how the land is exploited. In this context, it is interesting that during his survey of the archaeological ruins of Swat, Stein found what he interpreted to be evidence for a reservoir and irrigation works at the Buddhist stupa and monastic site of Tokar-dara, near to Bir-kot-ghundai, up in the hills (1929, 35). Use was apparently made of a spring on the hillside, but unfortunately Stein gives little description or detail of the structures.

In his study of the effects of partition on the irrigation schemes relating to the Indus River system, Michel (1967) looks at the impact of such major irrigation networks on the rural populations of Pakistan Punjab. He concludes that while the total amount of land given to agriculture in this area has increased greatly, the attendant increases in both subdivision and fragmentation of the land, and the huge increase in population has negated any immediate benefits to farmers themselves (ibid., 395-6). Further, he says that the major irrigation and agricultural development schemes of the 1950's and 1960's that involved the introduction of new seed varieties, fertilizers, irrigation and so forth, failed completely because they were imposed by the government, and so represented attempted change from the top down (ibid., 437-8). The original Swat River canal was completed in 1885, and irrigated an area of around 650 square kilometres in the Yusafzai plain to the north and east of the Kabul and Swat Rivers (Imperial Gazetteer of India 1904, 44, 120). The Malakand tunnel extension of the Swat River canal was proposed to provide irrigation to the area of the Yusufzai plain to the north towards the hills, and also in branch canals to supply Peshawar District and surrounding area. Stein describes the effect of the Upper Swat Canal as bringing fertility to the eastern half of the Yusufzai plain (1929, 10), but the 1897-8 Gazetteer gives a different interpretation of the new fertility in this region. The Gazatteer states that it "is also a question whether the former populous condition of the northern half of the district was not also due to the existence of canals" (Punjab Government 1897-8, 15), and then suggests that there is evidence for earlier canal systems in this area. The Kabul River canal, completed in 1893, provides irrigation for an area of approximately 200 square kilometres in the Peshawar and Naushera Districts (ibid., 44). The Kabul River canal apparently "is a revival of an old

Mughal canal" (ibid., 121), which is significant in terms of continuity of irrigation and agricultural intensity in this region, and may have an important bearing on deciding just how much really is new in terms of technology and its effects in this area.

3.6 Summary and conclusions

This account of the geography, land use and population in Dir, Swat and the Vale of Peshawar, (particularly Charsadda District), has highlighted a number of points. In terms of subsistence two major parameters appear to be altitude and water. Altitude dictates not only natural vegetation and animal life, but also affects climate, and so cultivation. Connected to altitude in all areas of the study (though not necessarily a direct result of it) is topography. In the low-lying Peshawar Basin, the nature of the land makes it ideal for intensive cultivation, although this can bring its own problems. In Dir and Swat, the further north, the higher the altitude and the more rugged and less suited to cultivation the land becomes.

Water, and in terms of cultivation, irrigation, is also a very important factor in land use. The greater dependence on irrigation in Charsadda District is understandable, as the average yearly rainfall here is between 250-370 millimetres, while in both lower Swat and lower Dir it is between 750-1000 millimetres. The use of the many rivers in the areas for both irrigation and as a source of fertility enhancement in the form of silt is an important part of cultivation management in this area. Different forms of irrigation and water management are important in Dir and Swat, as a result of their different natural aspects and requirements. The production of rice is far more significant in the Northern Valleys than in the Vale of Peshawar. In both Swat and Dir rice production is almost entirely by paddy, or irrigated methods.

While environmental factors are very important in dictating subsistence patterns, as can be seen from the preceding summary of data, other factors are important too. For example, the increasing number of buffalo in Swat could be attributed to changing tastes and demand for buffalo milk in areas where buffalo have not been so numerous until relatively recently. Sugar cane is now grown in great quantities in Charsadda District, and the vast majority of this is a cash crop, grown at the behest of the landowners (see Chapter 8). Both environmental and social or ideological parameters will be considered when examining the ethnographic and archaeological data.

The ethnic mix in this region is great, and the different accounts of these groups by different researchers, including those writing in similar periods, shows how poorly the groups and their subsistence patterns are understood. Some claim that Pathans were nomadic shepherds until recently, others maintain they have a long history of settled farming in NWFP. Gujars likewise, are alternately described as tenant agriculturalists, the true original inhabitants of the region, and cattle herding nomads. Although confusing, this situation does indicate just how complex the ethnic and related subsistence data are, and suggests that no single simplistic model is adequate to either describe or explain change and development in this region.

In terms of the social and physical changes in this region that are likely to have had some bearing on subsistence patterns, and so affect the relevance of the ethnographic data to the archaeological data, it is also difficult to reach any firm conclusions. The lack of established methodology for assessing the effects of such changes means that in part, this study is developing methodology as it progresses, and this may result in a lack of critical evaluation. Nevertheless, by using the combined approach of checking historical material for subsistence and land use information against both modern published data and the interview data, some control is gained. Any study of the proto-history and history of this region points to one major theme: that of change. From the rule by Achaemenids through to the Mughal irrigation schemes, it is clear that both political and physical changes have been occurring in this region, and both almost certainly affected in some degree the way the inhabitants have exploited and interacted with their land.

The area has been recognised in many accounts as fertile and ripe for intensification. It is likely, therefore, that schemes for increasing production have been carried out over many millennia, perhaps altering in terms of degree, depending on the political organisation and economic skill of the ruling powers. This is not to downplay the significance of recent events, which may be argued as being on a far greater scale than any before, but it should also be remembered that this region has seen great change for many centuries, and it is important not to underestimate the work and development of previous dynasties and rulers.

Nevertheless there is yet another theme of the basic integrity of the occupants. Caroe (1958) talks of the relative immunity of the hill people to political change, and while it is reasonable to suppose that the more inaccessible areas of NWFP would provide such natural protection, Swat and Dir, offered very important access and trade routes and are unlikely to have been entirely remote. When comparing the information from gazetteers compiled around 100 years ago, and the other historical accounts with the published agricultural census data, and the ethnographic interviews, continuity is very clear. Despite a century or so of major change in terms of politics and physical development, there are still many areas of great similarity and consistency.

Chapter 4
Northern Valleys Archaeology

The archaeology of selected occupation sites in the valleys of Dir and Swat is summarised here to allow an understanding of the social and cultural organisation of the site occupants, based on an analysis of the material remains, and in particular defines and summarises the chronology at these sites during the period between c. 1800 to 1200 BC. On the basis of widely divergent chronologies the two areas within the study, the Northern Valleys and the Vale of Peshawar, have traditionally been treated as largely separate and unconnected areas, each with their own main sources of influence and development (eg Dani 1967; Stacul 1987; Wheeler 1962). However, when the new chronology from the Bala Hisar of Charsadda (Ali et al. 1998; Coningham pers. comm., 2000) is presented in Chapter 5, it is clear that occupation here was contemporary with PIV occupation in the Northern Valleys. A number of social and economic theories and models of interpretation that have been developed by researchers in this region on the basis of the archaeological evidence that support the interpretation of the Northern Valleys and the Southern Plains having very little contact with each other during this period (Dani 1992, 1968; Stacul 1996, 1994a, 1994b; Tusa 1979). In Chapter 9 these theories and models are summarised, then critically evaluated in terms of how the archaeological evidence has been used to support them, and how relevant they are following a reassessment of the environmental evidence.

Seven sites, two from Dir and four from Swat, will be examined in turn, in terms of their location, material remains and the features and structures uncovered. These sites are located in the lower part of the valleys of Dir and Swat, in the northern section of the North West Frontier Province, and these sites and the chronological periods for which recovered material has been recorded are given in Table 4.1. Their position is shown on Figure 2. In total, more sites from this period have been excavated in Swat, than Dir, therefore, the sites selected from Swat for analysis and discussion here are those with published or available environmental data as this is very important in terms of the objectives of this research.

Aligrama, Bir-kot-ghundai, Ghaligai, and Loebanr III all have both faunal and floral material published, and Kalako-deray has had a faunal analysis published (Jawad 1998). The very limited environmental data from Dir is largely descriptive in nature, but in the absence of more detailed or recent work in this area, it was thought important to include archaeological data from the occupation site of Balambat, and where relevant, from the associated grave site of Timargarha (Dani 1967, 113, 239).

In line with the overall aims of this research the time period under examination is that from c. 1800 to 1000 BC, which is

Table 4.1 Occupation Sequence of Northern Valley Sites Period (Swat chronology)

Site	I 2970-2920 BC	II 2180 BC	III 1950-1920 BC	IV 1710-1300 BC	V 800-500 BC	VI/VII 500 BC +
Aligrama				x	x	x
Bir-kot-ghundai			x	x	x	
Ghaligai	x	x	x	x	x	
Kalako-deray			?	x	x	x
Loebanr III				x		
Balambat				x	x	x
Timargarha 3				x	x	x

sources: Dani 1967, Stacul 1969, 1987

Table 4.2 Summary Chronolgy of the Northern Valley Sites

Period (Swat Chron)	Site	^{14}C Date (cal)	Source	Interpretation
I	Ghaligai	2970-2920 BC	Stacul	
		2520-2230 BC	Possehl	Neolithic/Chalcolithic
II	Ghaligai	2180 BC	Stacul	
		1980-1870 BC	Possehl	late Harappan/early Kot Dijian
III	Ghaligai	1950-1920 BC	Stacul	
		1660-1560 BC	Possehl	Neolithic/Chalcolithic
IV	Aligrama	1360-1300 BC	Stacul	
		1710-1690 BC	Stacul	
		1210-1090 BC	Possehl	Chalcolithic
	Loebanr III	1730-1600 BC	Stacul	
		1560-1225 BC	Possehl	Chalcolithic
	Timargarha	15th-14th C BC	Dani	
		1590-1470 BC	Possehl	prehistoric necropolis
V	Aligrama	1540-655 BC	Possehl	protohistoric
	Timargarha	8th-9th C BC	Dani	Achaemenian
		1000-800 BC	Possehl	protohistoric necropolis

sources: Dani 1967, Possehl 1994, Stacul 1987
nb: Bir-kot-ghundai & Kalako-deray have been excluded from this table because no ^{14}C dates have been published

shown in Table 4.2, and covers all of Swat PIV, with some overlap with Swat PIII and PV. This period encompasses what is recorded as the shift from the Bronze to the Iron Age, and thus has been used in the culture-historical sense to describe ethnic groupings (Allchin 1995, 39; Dani 1967, 24). The relevance of this terminology has been discussed in Chapter 1, where its use was largely justified due to its continued recognition within Pakistan. This period has been selected as the focus for examining the archaeological material, because it is the period for which the environmental material, presented in Chapter 6, for this area has been recovered. It has also been interpreted as a significant period of cultural discontinuity, and the period during which the first substantial settlements appear in Swat (Stacul 1987, 1969). Grave sites belonging to the Gandharan Grave Cultural complex (Dani 1967, 24-30; 1992, 397) and the later Buddhist sites have largely been excluded, allowing a concentration on occupation sites of this period. These occupation sites may be connected with cemeteries, or overlain by or adjacent to later Buddhist material, but the study of both grave and Buddhist sites with the occupation evidence comprises an extensive and complex data set.

Given the importance of dating in linking the sites of Swat and Dir both to each other and to the wider region of South Asia, clarifying and condensing the existing chronology is essential. The results of combining the ^{14}C dates given by Stacul (1987), and Possehl (1994), the Swat Valley Chronology based on the Ghaligai sequence (Stacul 1969), and the subsequent relative dating of later sites on the basis of the Ghaligai work are presented in Table 4.2. The chronology constructed by Stacul known as the Ghaligai or Swat Valley Chronology (Stacul 1987, 1969) remains to date the single most comprehensive and important dating scheme for the late Neolithic and early proto-historic periods, and Gandharan Grave sites in Swat and Dir. From the original chronology established at Ghaligai (1969, 63), Stacul has used a combination of pottery, lithics and feature typologies to develop a relative chronology for the sites he has excavated in Swat. This has then been used to compare other sites in Swat and those excavated in Dir by the University of Peshawar team (Dani 1967).

One of the most important aspects of Stacul's chronology is that through the comparison of many sites, it has been used to detect trends in material culture and influences on this culture, both from within the region and from further afield. As the Swat chronology has been developed primarily from the work at Ghaligai, it is largely based on relative dating, and so rests on the uniformitarian assumptions that underpin such schemes. This is very important in terms of the fundamental archaeological questions being addressed by research in this area, such as the migration, diffusion and contact, as can be seen in the theories and models based on the archaeological material.

In order to determine whether the Late Bronze and Early Iron Age sites from this region do in fact form an associated group, each site will be briefly considered in terms of its material remains, and then a synthesis of the data will be presented. Existing interpretations that look at the sites as a group, and attempt to place them within wider developments in South Asia during this period will also be considered. This will include recurring themes in the interpretation of archaeology in this region such as isolation and operation as a closed system, trade and cultural links, and environmental determinism, and these will be explored through the published information about the sites and their artefacts.

As part of the distinction between rural and urban sites, which will be discussed further in Chapter 9, an attempt has been made to assess the size of each of the sites. While for the urban sites of the Bala Hisar of Charsadda and Taxila spatial measurements have been calculated and published, for the sites in the Northern Valleys, the situation is less clear. Here, the excavators have frequently quoted the dimensions of excavated trenches, but failed to give either measurements or estimates of the extent of the sites. Surface surveys in the form of recording or collecting pottery scatters are sometimes noted, but details such as area and quantity are lacking. Nevertheless, information does exist that gives a general idea of the extent of these sites, and this will be given where possible, to allow at least a comparison of scale between the settlements of the Northern Valleys and the Southern Lowlands. This is not intended as an exploration or re-analysis of the environmental data (this is dealt with elsewhere). Rather it is examining the way some of the data have been used in relation to the theories that have evolved to explain the archaeology of this region, and by looking at each phase across the whole region in terms of its homogeneity or heterogeneity at the seven sites, it is intended to highlight trends in both the data and in their interpretation.

4.1 Ghaligai, Swat

Excavations by the Italian Archaeological Mission in Pakistan were carried out at a rock shelter site near Ghaligai in 1967 and 1968. The area excavated, both within the rock shelter itself and immediately outside, exposed what the excavator described as a sequence of occupation from the second half of the 3rd millennium BC through to the second half of the 2nd millennium AD, the Islamic period (Stacul 1969, 44). Radiocarbon dates for the occupation give an earliest date of around 2,900 BC, which corresponds with Swat Chronology PI (Stacul 1987, 167), through to PVII, dated through pottery typologies to between ca. 500-400 BC (Stacul 1969, 85). The rock shelter is situated in the southern part of the valley of Swat, between the modern towns of Mingora and Barikot. In this lower part of the valley, the river has approximately two kilometres of flat, fertile land on either side, which in turn changes very suddenly into the steep hills that then lead into the mountains that separate Swat from the surrounding major valleys (Stacul 1967a, 185). Ghaligai is situated at the foot of a limestone cliff close to the modern passes of Malakand and Karakar (ibid.), and today the site is surrounded by cultivated fields. Ghaligai is in the Lower Swat area, where the River Swat has a wide surrounding fertile plain.

In 1967 work was carried out mainly inside but also immediately outside the rock shelter. The excavators have acknowledged some of the problems connected with archaeological work in a rock shelter site, such as ground water infiltration during excavation, rock falls outside the shelter that had penetrated the ground to a depth of three metres and, in nearly all the strata, pebbles and rocks fallen from the limestone rock face itself were recovered (Stacul 1967a, 187, 189-90). General problems associated with the excavation of rock shelter and cave sites include dating of deposits, and the possibilities of both multiple and sporadic use, and these are often compounded by the limited area of occupation material available (Branigan & Dearne 1992, 20). Difficulties associated with dating and the integrity of contexts in the excavation of rock shelters have been noted in the work at Franchthi Cave and adjacent sites in Greece (Wilkinson & Duhon 1990, 71). The dimensions of the cave prior to excavation were given as 8 metres wide at the opening, an interior depth of 7.5 metres, and a height of 1.8 metres (Stacul 1967a, 186). The subsequent excavated area in front of the cave yielding occupation material measured approximately 15 metres by 15 metres, reaching a depth of 12 metres (Stacul 1969, 47-8).

The earliest phase at Ghaligai (strata 24-21, PI, c. 2400-2100 BC; Stacul 1969) is characterised by stone tool manufacture and use, while occupation is in the shelter itself. The occupation evidence in the shelter takes the form of hearths, paved and levelled surfaces and possible postholes (Stacul 1967a, 185). Fine quality, wheel made pottery occurs in the next phase (strata 18-19, PII, c. 1810 BC; Stacul 1969), along with a limestone mortar (Stacul 1969, 53). In stratum 19 only, pottery of a distinctive vase shape in common with Harappa Cemetery R, Mundigak Period IV, Amri Level IIIA, and Kalibangan was recovered (ibid.). From strata 17 and 16 (PIII, c. 1505 BC; Stacul 1969), finds included pottery, such as hand made vases, some with mat impressions on their bases. The pottery of strata 17 and 16 is considered greatly inferior to that recovered from strata 19 and 18 (1967a, 208). It is noted that while pebbles and stones as artefacts and raw material for artefacts are absent in 19 and 18, they reappear here, alongside scrapers and flakers, rubbing and grinding stones (ibid., 54-6). The pottery from stratum 15 includes black, black-grey and buff burnished ware, and much of the black burnished ware has mat impressions on the base, which is argued as evidence for cultural continuity between strata 17 and 16 and stratum 15 (Stacul 1969, 62).

Stratum 15 is defined as belonging to both PIV and PV of the chronology (1969, 83-4). PV is considered contemporary with the archaic phase of the proto-historic graves of Swat, the graves of Timargarha in Dir, and pottery from the deepest levels at Charsadda (ibid., 84). Stacul's radiocarbon samples from the 1967 and 1969 excavations give dates between 1500 and 1000 BC (ibid.). Copper objects were first recovered from Ghaligai in stratum 15, or PIV layers (ibid., 61), and iron from the later periods (ibid., 81; 1967a, 214).

4.1.1 Interpretation of Ghaligai

Stacul says that the early evidence for a stone tool industry at Ghaligai combined with the animal remains of antlers and boar tusks "show economic activity related to the traditions of hunting peoples. It is also possible that such human groups lived on the margins of other communities (agricultural?) with already established settlements" (Stacul 1969, 50-1). Yet a shouldered hoe was recovered from strata 24-21, the earliest layers (1969, 49), which suggests not only the utilisation of plant food, but an agricultural component. While the limestone mortar from strata 18 and 19 (ibid., 53) may show either agricultural processing, or be related to the processing of gathered plant food, a hoe is much stronger evidence of cultivation.

Strata 17 and 16, Swat PIII-IV, contrast greatly with the earlier strata 18 and 19 as much less 'refined'. "Such a radical change from the preceding horizon can obviously not be explained away as due to merely occasional influences. It should more likely be related to the brusque superimposition of a new culture, far different in origin from the preceding one, whose appearance may perhaps be connected with extremely significant migratory phenomena" (ibid., 56). The similarities in pottery styles with sites such as Mundigak in Afghanistan, also led Stacul to suggest other similarities in the cultural development such as the "origins of the 'barbaric' culture which characterized the perhaps contemporary phases of Ghaligai (strata 17 and 16), Burzahom (Period I) and Mundigak (Period V), should probably be sought for in the north" (Stacul 1967a, 210). This includes the occurrence of dwelling pits at Burzahom. From the later layers at Ghaligai, Stacul notes similarities between the pottery of stratum 13, PV, and that from levels 38, 37 and 35 at Charsadda (1967a, 213).

Stacul says that typologically there are very clear similarities between layer 15 at Ghaligai and the deepest occupation level at Bir-kot-ghundai (Period IV, c. 1700-1400 BC) and at Loebanr III (Period IV) (1969, 62). The changes in the standard of pottery between the phases or periods at Ghaligai is, according to Stacul, indicative of a great deal of discontinuity (1969, 85). What does this mean in terms of culture change? Stacul's summing up within the interpretation of Ghaligai is that these sites during this period "were exposed…to a prolonged renewal of cultural waves coming from without, accompanied by sudden upheavals with their relative widespread phenomena of immigration on a large scale. This climate of instability makes defensive needs predominate, and leads communities to settle in hill zones of difficult access, as shown by some of the areas in which remains of this culture were found" (Stacul 1969, 64). Stacul also says that it is possible that the abrupt changes in cultural indicators between strata 19 and 18 and then strata 17 and 16 may be linked to the invasion and spread of Indo-Europeans across Eurasia (1969, 56-7).

4.2 Aligrama, Swat

The Italian Archaeological Mission of IsMEO, under the directorship of Professor Facenna, has undertaken extensive exploration and excavation in the Swat Valley, and this included excavations at the site of Aligrama in 1966, 1972, 1973 and 1974 (Stacul & Tusa 1975, 291; 1977, 151). The modern village of Aligrama is six kilometres north west of Mingora, situated in the valley of the Shandheri Khwar, a tributary of the main Swat river (ibid.). The site itself is located some 100 metres above the valley floor. Aligrama is also situated within Lower Swat, and so would have had access to fertile land and an adequate supply of rainwater for grazing and the production of crops.

The earliest occupation evidence at Aligrama consists of a compacted soil layer, and the earliest construction phase is represented by a number of dry stone walls that enclose a series of linking rooms (1975, 294, 301). It is also suggested that four pits belong to this phase, however, "it is difficult to say for certain since there are no connections with the beaten earth floor of room 3 and, moreover, there is a complete lack of stratigraphic connection" (ibid., 302). The earliest construction phase has been dated, on the basis of artefact typology, to PIV in the Swat Chronology (Stacul & Tusa 1977, 159). A number of trenches were excavated at Aligrama, and these were placed in an area approximately 325 metres by 400 metres, adjacent to the modern village (ibid., 152). Although the excavators did not explicitly discuss the size of the occupation area in their reports, the placement of the trenches suggests that they expected it to have been relatively significant.

Five later construction phases were assigned to PIV (ibid., 161), and much of the evidence for these later phases consisted of extensions to the existing walls, quite roughly built (Stacul & Tusa 1975, 302-3; 1977, 157-8), and fireplaces, one of which was lined with schist slabs, dug into the floor. In the corner of one of the rooms, a line of stones separates an area which the authors say may have been used for the storage of vegetables, but without offering any supporting evidence for this specific interpretation (Stacul & Tusa 1975, 304). The latest phase is also represented by stone walled structures, but there are some changes to the wall placement and related features (ibid., 305), mainly a decrease in numbers of fireplaces and the inhabited area, which is argued represents a decrease in population at the site (ibid.). All the walls are built in a dry-stone technique, from river pebbles or stones, and the authors say that this uniformity in material and technique precludes more precise phasing of the structures (ibid., 294). Stacul and Tusa suggest that the pits uncovered at Aligrama are for the storage of "food-stuffs or other things" (ibid., 298).

The pottery from Aligrama recovered in the first two seasons of excavation was used to give a relative date to the earliest phases of occupation, as it is claimed that the pottery types are directly comparable with those recovered from the earliest phases of the proto-historic graveyards at Loebanr, Katelai and Butkara II (ibid., 310). This would place Aligrama's earliest occupation in PV of the Swat chronology (ibid., 320). However, following the pottery recovered from later excavations, the presence of gritty grey and brown pots, black-grey burnished ware and red or brown ware from the earliest occupation layers, a date from PIV was given (Stacul & Tusa 1977, 159). A number of miniature vase was recorded from the PIV layers (Stacul & Tusa 1975, 314). Both copper and iron objects were recovered during the earlier excavations at Aligrama, and while the copper objects range from PIV layers through to later periods, the iron objects appear to have been collected from layers that are considered on relative dating grounds to be from the middle periods of the graveyards of Loebanr, Katelai and the other Swat sites, which equate to PVI in the Swat chronology (ibid., 317, 319-20).

At least six grindstones, one mortar and a pestle were recovered at Aligrama, which suggests that plant food processing was taking place on a regular basis here (ibid., 309). Whether these finds are connected with agricultural production, or whether they are associated with the preparation of gathered plants, or even plants for a non-food purpose is rather difficult to ascertain without analysis of a sample of plant macro-remains from the site.

4.2.1 Interpretation of Aligrama

The structural phases at Aligrama are believed to show an increase in population, followed by a decrease (Stacul & Tusa 1975, 308). The number of fireplaces and the presence of miniature vases together somehow produce an interpretation of the use of these vases in the production of foodstuffs associated with dairying, and thus the occupants of Aligrama at this time were, according to Stacul and Tusa, obviously pastoralists (ibid., 309). They also continue to say that this is supported by the almost complete lack of animal bones, which "may be due to chance in the documentation which has come down to us or to an actual distinctive element in the people who inhabited the site. This latter hypothesis would not be completely in contrast with the physiognomy of a settlement of herdsmen - dairymen, which we deduced from the few elements at our disposal" (ibid., 309). Yet the authors hesitate to pronounce upon the significance of the grindstones as indicators of agricultural activity at the sites, or even wild plant processing (ibid.).

The pits uncovered at Aligrama are not interpreted as dwelling pits (Stacul & Tusa 1977, 158), although they are included in construction phases. This interpretation may be because the pits are not part of PIV, but rather PV, which falls outside the main period of influence from Kashmir and the Northern Neolithic according to Stacul. This means that Aligrama is a Swat site, with clear occupation evidence from PIV, assigned on the same pottery and artefact typology, yet it has stone wall structures that are not preceded by a phase of pit dwellings. When compared to the evidence for pit dwellings preceding dry stone wall structures from the other Swat sites, this suggests a new departure with regard to the interpretation of the pits, and so the possible occupation sequence.

The extensive material from the later periods at Aligrama, and their similarity to material from Balambat and Timargarha in Dir, and Charsadda and Zarif Karuna from the northern Vale of Peshawar, have led Stacul to suggest "the presence of a 'cultural province' which, in spite of local variations, is relatively homogeneous" (1977, 175). That Stacul is discussing material and comparisons which he places in PV-VII of the proto-historic chronology, or those equivalent to the grave culture, is largely due to the constraints imposed by his and other chronologies (1977, 177). It is hoped that the analysis and re-interpretation of the environmental material, and the inclusion of the ethnographic material will lead to a new model of contact and change, showing that this type of region-wide contact was not new in PV, or even confined to the duration of the Gandharan Grave Culture.

4.3 Bir-kot-ghundai, Swat

The Italian Archaeological Mission excavated the site of Bir-kot-ghundai, in 1977 and 1978 (Stacul 1980a, 55; 1978, 137), and then returned for a second phase of excavation between 1985 and the late 1990s, concentrating on the Historic Periods (Callieri 1992, 339). Bir-kot-ghundai is located in the main Swat Valley, close to the modern village of Barikot, and Plate 3 (page 372) shows part of the recently excavated area. The site is situated on the lower slopes of a hill that reaches 943 metres, and excavation was carried out on the lower, terraced slopes of this hill (Stacul 1987, 53, 60). The site is close to the valley floor, and hence the Swat River, at one of the widest points of the valley itself. In addition to occupation evidence, there is a proto-historic graveyard associated with the site (Stacul 1978, 137). Estimates of the area of pre- and proto-historic occupation at Bir-kot-ghundai are not given, and the excavation reports published by Stacul (1980a, 55; 1978, 139) give only the trench dimensions. In 1978, one of these trenches reached a size of 42 square metres, and there is also known to be a great deal of multi-period material over the hillside above the modern village of Barikot (Stacul 1980a, 55; 1978, 137).

The features excavated comprise pits and walls. The walls were made of irregularly shaped stones and river pebbles. There were also clay floors, some with stone slabs fitted into them, ash and charcoal deposits, pot sherds, animal bones, and terracotta figurines, and stone and bone tools (Stacul 1980a, 56; 1978, 138-9, 147). Earliest occupation at Bir-kot-ghundai has been dated to Ghaligai/Swat chronology Period IV (c. 1700-1400 BC), and the structures from this period are both pits and walled rooms (Stacul 1987, 61; 1978, 137). However, there are believed to be two phases in PIV here. The 'semi-subterranean' structures, or the largest pits, with some material that is interpreted as occupation evidence in their fill or construction, indicate the earlier phase, and the above-ground walled rooms indicate the later phase (Stacul 1987, 61; 1980a, 61). Determining the relative relationship of these features to each other, and so their respective dates, is based largely on dating the finds associated with the features. These finds are assigned on the basis of their typology to the chronological sequence from Ghaligai, or the Swat sequence (1980a, 57). The pits at Bir-kot-ghundai are not really discussed in great detail in the excavation reports. Rather, the similarity to the structural sequence at Loebanr III is noted, and the same interpretation implied (Stacul 1980a, 61).

During the later excavations, what has been interpreted as a fortification wall around at least part of the proto-historic settlement has been identified (Callieri 1992, 339). A construction date of 3rd century BC has been suggested for the fortification wall and bastions, based on the relative dating of both human and animal terracotta figurines, and some distinctive sherds. This material has been compared to finds recovered from the Bala Hisar at Charsadda, and assigned dates of the early 3rd century BC (ibid., 343). Callieri proposes a link between the fortifications at Bir-kot-ghundai and the eastward expansion of the Greeks, especially the identification of particular sites in the Northern Valleys with the conquests of Alexander the Great (ibid., 343-5). However, the new radiocarbon dates calculated for the Bala Hisar of Charsadda (see Chapter 5) allow a major re-assessment of the chronology of this area, and if the defenses of Bir-kot-ghundai have been given an early 3rd century BC date on the basis of artefact comparison, then this is likely to be affected.

Evidence for a major defensive construction in the form of wall and bastions is significant. As Callieri notes "a fortification wall is a major building in the history of a town, implying a large economic effort or a strong political will" (ibid., 343), and suggests that Bir-kot-ghundai was an important centre, and although no estimation for the length of the wall is given, it is likely to have been a town of some size. The role of the defensive ditch at Charsadda will be discussed in the next chapter, as it is possible to suggest more than one reason for building such a structure.

The pottery from PIV has been classified into three major types. These are grey-brown gritty ware, black grey and buff ware, burnished red and black, and black on red painted ware. Stacul says that these types are the same as those recovered from Ghaligai, Loebanr III and Aligrama (1978, 140). Indeed, the comparable layers from Aligrama and Loebanr III have been dated by bristlecone pine dendrochronology to between 1700 and 1500 BC (ibid., 149). Within the painted red ware, decoration of zebu were noted (Stacul 1978, 147), and the terracotta figurines also included at least three humped cattle figures. The PIV pottery also includes some specimens with mat or basket impressions on their bases (1980a, 58). Miniature vases were also recovered from layer 4, and through typological comparisons with artefacts from graves, these have been assigned to PVII (1978, 148). A quernstone was also noted in PIV, from a floor layer (1980a, 56) which may indicate agricultural processing. Copper artefacts have been recovered from PIV, along with gold (1980a, 59; 1978, 147), and those from PVII are said to 'include' three copper items, although it is not made clear whether there is a larger, undescribed sample (1980a, 61).

4.3.1 Interpretation of Bir-kot-ghundai

The earliest occupation at Bir-kot-ghundai is assigned to PIV of the Swat chronology, and comprises two structural phases. The earlier is thought to be represented by the pits cut into the soil and the later by the stone walls, and Stacul (1980, 62) notes the similarity between the phasing and structures of Bir-kot-ghundai and Loebanr III. In turn, the use of subterranean or pit dwellings have been linked with the influence of Burzahom, and other sites to the north. Other influences from outside Swat are seen in the presence of the painted black on red ware, of which Stacul says that "particular subjects and motifs of the painted decoration suggest generic affinities with decorative patterns common in the western region of the sub-continent during the Indus Civilisation period and later" (1978, 150). The painted black on red ware disappears from Swat after Period IV.

The similarities between Bir-kot-ghundai and the other proto-historic occupation sites in Swat, both in terms of the structures and the pottery, are noted (Stacul 1980, 61; 1978, 149). PIV has been emphasised, but the pottery of PV at Bir-kot-ghundai was considered to be very similar to the material from Aligrama (1980, 64). Up to PV, the only metal artefacts recovered were copper and gold (1978, 147; 1980, 59), but from PVII, the occurrence of iron artefacts and weapons in graves are noted (1978, 150).

4.4 Kalako-deray, Swat

Kalako-deray is a flat hill top, covering an area of c. 85 by 35 metres, and is 1130 metres above sea level (Stacul 1993, 69; 1995, 109). The hill itself is near the modern village of Kukari, which is 10 kilometres east of Mingora, and Plate 5 (page 373), shows part of the hill top with buildings encroaching, as it was in 1998. Kalako-deray rises above the stream of Narkat Tange that is a tributary of the Jambil Khwar (Stacul 1993, 69). The height of Kalako-deray places it in an area of temperate coniferous forest, in contrast to the other Swat sites (Roberts 1997, 9-11), and the narrowness and steepness of the side valley also limits the amount of available fertile land. A joint Italian - Pakistan team excavated the site in 1989, 1990, 1991, 1992, 1993, 1994 and 1996. Today, the hill top has been ploughed for cultivation, and although the upper, steep slopes of the hill are not under cultivation, the lower slopes are farmed and have been built on. The area of the hill top where the excavations have been focused was given as approximately 2000 square metres (Stacul 1993, 69), and although there is at least one farmhouse on the slopes of the hill today, the hill itself is steep, and likely to have discouraged settlement.

In the first excavation report, Stacul says that the earliest occupation at Kalako-deray is from Swat chronology PIII, dated to the beginning of the 2nd millennium BC, and this is represented by a floor level and associated occupation material, although only a small part of this was uncovered. This underlies a number of pits from the later PIV (1993, 69, 72). However, in the later excavation reports, Stacul says that Period IV, represented by a number of pits in a range of sizes, shapes and attributed functions, is the earliest occupation period of the site (1997, 363; 1995, 109). Period IV is dated to between the 18th to 15th centuries BC, and following a very long gap in evidence for occupation are PVII and PVIII, which are dated to around the 5th and 4th centuries BC (ibid. 109; 1993, 73-4). With the exception of the early floor level and later, historic stones walls, the occupation features from Kalako-deray comprise pits from the proto-historic periods (Stacul 1997b, 365; 1995, 110-15; 1993, 88). Although no post-holes have been excavated, fragments of daub with wattle or reed impressions that are thought to be part of a roof were found in association with two pits from PIV (Stacul 1997b, 375; 1995, 124).

The early pits of PIV were both circular/oval and square in shape, with the largest of these measuring 4.5 metres wide by 3.5 metres deep (Stacul 1993, 88). The stone artefacts and plant and animal remains recovered from the bottom of these pits is interpreted as indicating intensive occupation within them, and this phase at Kalako-deray is compared to the large settlements of semi-subterranean structures from both Loebanr III and Bir-kot-ghundai (ibid., 88-9).

The pottery recovered from the elusive Period III was gritty brown hand-made ware, some pieces with mat impressions on the base, and this was followed in PIV by typical examples of Swat Period IV pottery. This included wheel-made gritty brown and grey ware, often burnished, hand made miniature pots, rippled rim ware, pots with mat impressions on their bases and a very few sherds of painted black on red ware (Stacul 1993, 78; 1995, 118). In Pit B7 miniature vases and a burnished black bottle were recovered from beneath a paved floor (1995, 111, 124). Copper objects only were recovered from the PIV layers at Kalako-deray, while iron objects were recovered from the later pits and layers (1995, 123; 1993, 86).

Kalako-deray is unique in Swat during this period for the number of sickles which have been recovered. In one season alone, 13 holed sickles were found, and along with 36 grinding slabs and seven mortars are believed to show how important agriculture was at Kalako-deray (Stacul 1994b, 238), in particular during PIV, as the number of recovered heavy ground stone items from PVII and PVIII is by comparison, very small (Stacul 1995, 124). Previously, only a very few agricultural stone tools had been found at other Swat sites from PIV (Stacul 1993, 89; 1995, 124). The recovery of one double holed sickle, which is similar to those from the Kashmir Neolithic, is interpreted as significant in terms of influence from the north, or 'Inner Asia' (Stacul 1995, 124). Where quern stones and other stone agricultural implements were found within the fill of pits, Stacul has suggested that the pit itself was used as an area for milling (1993, 91).

4.4.1 Interpretation of Kalako-deray

Notable at Kalako-deray is the long time span separating both PIII from PIV, and then PIV from the later PVII and VIII (Stacul 1993, 73, 88). This is explained in terms of PVII, at Kalako-deray and other Swat sites, representing a completely new and different type of cultural occupation of the area, that has no evident links to earlier phases (ibid., 91). The presence of iron objects and the number of flat human figurines in PVII, as well as new pottery styles (Stacul 1995, 124) confirm this. Two possible explanations are given: "(1) Period VII represented an internal evolution from the previous Swat Period VI; the evidence of the new occupied sites in the peripheral hilly area, including Buner, should be connected with population growth and an increase in agricultural activities. (2) Period VII represented a new culture, which promoted an integration process in the main Valley and the occupation of new sites in surrounding area" (Stacul 1993, 91). In contrast, the development of PVII to PVIII is one of continuity and gradual evolution in pottery and other artefact styles (Stacul 1995, 124; 1993, 91).

The presence of what Stacul interprets as dwelling pits, notched and holed sickles, holed axes and jade beads from Kalako-deray (and some of the above from other Swat sites) are seen as evidence of links with what is called the 'Inner Asian' or 'Northern Neolithic' traditions, with comparable sites such as Burzahom (Stacul 1993, 89). Further, "Such 'northern' connections have not only been attributed to long-distance trading or infiltration of peoples, but also to a self-supporting cultural area around the great mountain chain" (ibid., 90). This idea will be explored in greater depth in Chapter 9, which examines models of social organisation in this region. With regard to the varying function of the pits at Kalako-deray, the presence of the zebu figurines and the miniature vases and bottles, along with a pair of muntjak horns from the base of another very small pit (without other faunal remains), has led Stacul to suggest a possible ritual, or non-utilitarian purpose for some of the pits (Stacul 1995, 124).

4.5 Loebanr III, Swat

Loebanr III is situated in the Jambil Valley, running east from the main Swat Valley, and four kilometres east of Mingora (Stacul 1987, 55). It is approximately 150 metres from the Jambil Khwar, up a steeply sloping hillside. The proto-historic grave sites of Loebanr I, Loebanr II and Butkara II are found 500 metres to the east, 300 metres to the west, and one kilometre to the west of the occupation site respectively. The occupation site is located on the flat top of a fan-like mound, with steep sides (Stacul 1987, 55), which measures approximately 140 metres by 110 metres, and it is suggested that it is here that the occupation was concentrated (Stacul 1977, 228). Today, to the south of the site, the hillside has been terraced and cultivation is carried out where the absence of rock permits (Stacul 1976, 13). Loebanr III is thus still within Lower Swat, and its location in a relatively broad valley means that access to fertile land was likely. The Italian Archaeological Mission in Pakistan carried out excavations at Loebanr III under the direction of Professor Stacul in 1968, 1976 and 1979 (Stacul 1976, 1977, 1980b).

The features recorded at Loebanr III comprise pits, in a variety of sizes and shapes, and walls constructed from river pebbles. Both are dated to PIV in the Swat Chronology, with the pits earlier in the period, and the walls which overly the pits, constructed towards the end of the period (1977, 229). The pits, or what Stacul calls the subterranean or semi-subterranean structures include a large ditch and connected pit. Within the fill of (presumably) both the ditch and pit were found sherds, animal bones and charcoal. Hearths and charcoal were also found on the south side of the surround of the ditch. On the bottom of the pit many irregular, shallow holes were uncovered. In relation to the ditch Stacul says that "Near its north side four deep post-holes, connected with traces of probable trellis or reed grating, were distinguished in the natural soil" (1976, 16). Further, in relation to the pit he suggests that there are irregularly shaped 'potsholes' around the outside near the opening. Stacul also says that clay with trellis impressions were also recovered during excavation, probably the result of collapsed building, the implication being that this was a shelter or extension of the pit/ditch. The sizes of the hearths outside the ditch compared to those inside suggest to Stacul that the greatest fires were built outside the ditch.

The pottery from Loebanr III is made up of brown to grey gritty ware, black-grey and buff burnished ware, red ware and brown ware (1980b, 70-1; 1976, 17), and mat or basket impressions on the base of pieces are common (1977, 245; 1976, 24-5). Design elements that have helped place the site of Loebanr III within the pottery typology for the Swat Valley include rippled or notched rims, new in PIV (1977, 245).

Other finds at Loebanr III include a stone quern near a hearth in the bottom of Pit H4/H7 (1980b, 73), and a stone knife (ibid., 74). The latter is linked by Stacul to similar knives recovered in Kashmir, and as these are attributed to influence from China and the north, this is seen to carry through to Swat. A jade object here is also seen as confirmation of these northern connections, and also the discontinuity between periods in Swat. Jade has been recovered only from PIV layers at sites, and not from other later phases (1977, 251). Only three metal objects have been recorded from Loebanr III, a fragment of copper wire and an iron arrow-head, both from PIV (1977, 250), and a copper chisel from the upper layers in the post-pit phase (1980b, 72).

4.5.1 Interpretation of Loebanr III

Stacul says that Loebanr III has helped show some of the more typical aspects of life and the settlement patterns in Swat. He summarises this saying "the people settled there were essentially sedentary, as confirmed by the deep pits cut in the ground, as well as by the wealth of fragments of very large vases and finally by the particular type of cattle-breeding" (1976, 30). This contrasts to the interpretation of the miniature vases recovered from Aligrama. Here, over 79 miniature vases similar to those recovered from graves at Loebanr II, Katelai and Butkara II were found (Stacul & Tusa 1975, 314), and these were interpreted as indicating that the occupants of the site must have relied primarily on herding. The miniature vases apparently were crucial to the processing of milk products (ibid. 309). The many pits excavated at Loebanr III have been assigned a number of functions, depending on their size and shape. These functions include dwelling pits (Stacul 1977, 250), storage, as fire places and possibly even as a kiln for copper casting (Stacul 1987, 56; 1976, 16).

From the 1979 excavations, Stacul says that the features and finds confirm that the underground structures relate to Period IV in the Ghaligai sequence, c. 1700-1400 BC (1980b, 72). He also says that the shape and form of the pits varies in relation to their position on the site, i.e. the natural and other topographical features, but also that differences in the depth, shape and size of the pits means that they were dug for different purposes (ibid., 72-3). With respect to Pits H4 and H7, where finds in the fill include ashes mixed with zebu terracottas, and a sherd with zebu graffito, Stacul suggests that the use of hearths connected to the finds means some type of cult ceremony was taking place.

A change in settlement is shown by the abandonment and sealing of the pits and the erection of stone walls over them (ibid., 74). The spread of black-burnished pottery, bone tools and metallurgy is noted at both Loebanr III and Bir-kot-ghundai at the end of PIV (c. 1500-1400 BC). This is thought to indicate a fundamental cultural change that started at the beginning of PIV. With the simultaneous expansion of site sizes to accommodate increasing populations (which is not quantified in any of the site reports), Stacul goes on to say that this expansion in population is likely to have been accompanied by the development of settled agriculture, shown by the type and variety of floral remains known from Loebanr III (ibid., 75-6). He explains the change in structures from pits to above-ground rooms during PIV as the result of contact with different groups. The relative slowness of the change in structures compared to the change in artefact styles he explains by saying "The change in this pattern of residence may have been finally produced by the recorded contacts with peoples of the great southern plains, but it seems originally connected with the great cultural change which started in the valley around 1800/1700 BC at the beginning of this culture" (Stacul 1980b, 76).

4.6 Timargarha 1-3 and Balambat, Dir

The Gandharan Grave Culture site of Timargarha is located on the east bank of the Panchkora River within the modern town of Timargarha, just above a flood bank of the river itself (Dani 1967, 59). An unnamed khwar runs close to the site, and there are hill ranges on all four sides (ibid. 60). The underlying rock here is granite, evident in a series of outcrops, but where there is sufficient soil cover, cultivation is carried out on either terraced slopes or the silt bank of the Panchkora (ibid.). Timargarha and Balambat are situated in Lower Dir.

The main Timargarha grave site is at the base of one of the many rocky ridges, and was excavated by a team from the University of Peshawar, led by Professor A H Dani in 1964 and 1965 (ibid.). Dani believes that the Gandharan Grave Culture is the culture that spans the period between the end of the Indus Civilisation and the start of the historic period with the coming of the Achaemenids in the 6th century BC (ibid., 8). He also claims that there are two waves of migration or invasion and settlement apparent in the graves and their associated artefacts. The first of these two waves occurs in the second quarter of the 2nd millennium BC and is characterised by bronze, and the second wave occurs around the beginning of the 1st millennium BC and is characterised by iron goods (ibid., 42, 51). In the analysis of metal artefacts from the graves and the occupation sites, it appears that copper objects alone are found in the PII layers, which are dated to around the end of the 2nd millennium BC, and this is followed in PIII, dated to the beginning and first quarter of the 1st millennium BC, by layers in which both copper and iron objects were found - a transitional period. In PIV, described as the period up to and including the beginning of the Achaemenian period, iron objects clearly dominate (ibid., 201, 239-40).

The pottery from the graves at Timargarha 1 were divided into two main groups of red ware and grey ware (ibid., 121), including what appear to be rippled rim type vessels (ibid., 127). With regard to the pottery from Balambat and Timargarha 3, Dani reports that the pottery types are very similar to those found at Timargarha 1, with some new types and sub-types, but that there are no vessels or sherds from period III here. Rippled rim pieces were also recovered from these sites (ibid., 249-50).

The oldest graves at Timargarha have been given a date of around 1500 BC on the basis of a single calibrated radiocarbon date (Dani 1967, 37; Possehl 1994, 116). These graves fall into Timargarha Period I, while the second radiocarbon date from a burial attributed to Timargarha Period III places it around 940 BC (ibid.). The date for the earliest burial phase is equivalent to Swat PIV, and is therefore earlier than the Gandharan Graves from Swat, which are dated to Swat PV, see Table 4.2 above.

The excavations at the main Timargarha site were extended across the valley to the west bank of the Panchkora to the site of Timargarha 3, which was situated on sloping land below the modern Balambat fort, on the highest point of the granite outcrop. During the excavation of this particular site no burials were found, but there were 20 pits dug in what Dani describes as a 'haphazard' fashion (1967, 113, 237). The fill of these pits was a heterogeneous collection of bones, charcoal, ash, sherds and other material, and this has led to them being interpreted as refuse pits, probably for the occupation site of Balambat itself (ibid., 178). Due to the nature of the pit fills it "was therefore not possible to take them as sacrificial pits" (ibid., 113). The pottery from Timargarha 1 has been assigned the same classifications and phasing as that from the main site, and is mainly from Timargarha Period II, placing it around the last quarter of the 2nd millennium BC (ibid., 182).

The occupation site of Balambat is adjacent to Timargarha 3, and four different structural periods were recorded at Balambat. The occupation site of Balambat as it was in 1997 is shown in Plate 7 (page 374), where some stone walls are still visible. The earliest remains are from Timargarha Period II, and comprise disturbed graves under later walls (ibid., 239). Period III consists of fragmentary stone walls (overlying the graves) with stone lined pits dug among them, iron and copper objects, stone tools and terracotta human figurines (ibid.). Period IV has completely new alignments for stone walled buildings, over those of Period III, and there are fire places in every room (ibid.). Timargarha Period IV is considered to be Achaemenid, and is the last building phase (ibid., 240). There is then a gap in the occupation of Balambat, from the end of Period IV until PV, which is thought to stretch from the Kushan to Hindu Shahi periods.

The walls of the Period II buildings are of stone rubble masonry, but had been badly robbed. Built into the walls are circular or rectangular store rooms, lined with rubble stone masonry, and within the rooms themselves, pits had been dug, filled with a variety of material, including animal bones, pottery and two terracotta bulls (ibid., 242-3). The excavation uncovered eleven rooms from Period IV, built of stone in what is described as a rough style of diaper masonry, and a number of burnt logs from within the rooms have been suggested as possible roofing material (ibid., 244). Pits were also dug within these rooms, and the fills of charcoal and ash have led to the suggestion that they were associated with the fire altars and fire places in each room. Complete jars and a grindstone were buried under a floor surface, and Dani suggests that the grindstone was used for medicinal herbs (ibid., 246). Rahman identifies a 'gardening' tool and a 'sheep shearer' among the iron objects from Period IV (1967, 275), and also stresses that "No less important is the occurrence at Balambat of pointed-butt ground stone axes of the Neolithic type commonly known in South India" (ibid., 284). Again, no details about the estimated size of the occupation area are given for either Balambat or Timargarha, but the excavated trenches are given as approximately 110 metres by 60 metres at Timargarha 1 in 1964, and 30 metres by 25 metres at Balambat in 1965 (Dani 1967, 61 & facing page 241, fig 48).

4.6.1 Interpretation of Timargarha 1-3 and Balambat

Timargarha and Balambat are considered to be part of the Gandharan Grave Culture, and span both the Bronze and the Iron Age (Dani 1967, 41-2). The date from the early grave at Timargarha of c. 1500 BC is interesting in terms of the dating of the Gandharan Grave Culture and its spread throughout the northern valley areas. As the main concentration in both excavation and report has been on the graves, and the earlier levels which may have related to Swat IV and earlier have been neglected, a true characterisation of the settlement sites is not possible. This has been exacerbated by events such as that described by Dani at "Balambat we observed some pit circles, but before we could excavate them, they were dug up by the local villagers" (1967, 9), with the implication that these may have been early structures associated with Neolithic occupation of the area. However, links between sites in the two valleys were also noted by Stacul following his excavation at Aligrama in 1974, when he made a specific comparison between Aligrama and the Peshawar excavations in Dir (1977, 174). Stacul made a connection between the cultural and the physical similarities of the two valleys, and suggested that due to the geographic and climatic homogeneity, cultural homogeneity was natural (ibid.).

Of the settlement excavation, it can be seen that the presence of numerous pits of different shapes and sizes is in keeping with many of the Swat sites, although at Balambat the pits seem to be stratigraphically linked with the stone walled buildings of both PIII and PIV. The recovery of pottery from PIV at Balambat has produced forms which equate with the so-called Achaemenian period (6th-4th centuries BC), and are similar to forms recovered from equivalent layers at Charsadda. With respect to the ground stone tools, in the form of ring-stones or mace heads and pointed butt stone axes, Dani compares them to tools discovered by Marshall at Taxila and at Serai Khola near Taxila (1967, 285-6). They are also likened to those from Neolithic Burzahom, and the similarity with both northern and southern artefacts gives rise to Dani's suggestion that the ground stone tool manufacturers and users migrated south into central and southern India (ibid., 287). The recovery of two terracotta bulls from PIII is another similarity with the Swat sites. The lack of plant or animal remains from Balambat is likely to be as much a product of the sampling policy (or lack of it) as a real outcome. The tangential evidence for subsistence, such as the gardening and shearing tools and the grindstone will be used in Chapter 6 to support the primary material from the Swat sites.

4.7 General interpretation

In terms of determining whether these sites are rural or urban in nature, there seems to be little in the way of structural or artefactual evidence that suggests they were cities of the magnitude of the Bala Hisar or Taxila during the period under discussion in this study. Bir-kot-ghundai alone has an increase in size, and the presence of a fortification wall in phases that have been assigned to the latter half of the 1st millennium BC (Callieri 1992), and from this, may be considered an incipient urban site. Nevertheless, during PIV (c. 1700 - 1400 BC) these sites were obviously important settlements within the Northern Valleys, and their situation on trade routes is likely to have contributed to this (see Figure 8, page 272). That they do not fulfil all, or indeed many, of the criteria designated necessary to be classified as a major urban site, does not prevent them from having been significant to a mixed sedentary, seasonally mobile and transitory population.

The internal coherence of the pottery sequence and the other artefacts' typologies has allowed the development of an overall chronology for the Swat valley, into which sites such as Timargarha and Balambat can also be placed. This chronology was developed from the early excavation at Ghaligai, and supported by radiometric dates from this site and others in Swat. Possehl's re-examination of radiometric dates (1994) has changed some of the boundaries, and a re-assessment of the periods and dating of the Northern Valley archaeology will be developed as part of the new model, once the material from the Vale of Peshawar has been presented. It can be seen from Table 4.1. that the only period for which data from all of the sites considered in this chapter has been recovered, is PIV. As this is the period from which the majority of environmental data have been collected, this period necessarily becomes the focus for the archaeological summary.

Similarities can be seen in, for example, the pottery, which at all the sites can be broadly categorised as black-grey and buff burnished ware, brown-grey gritty ware, and red and black on red painted ware (Stacul 1987, 79). Differences between sites are noted in the relative quantities of different types, for example red ware at Aligrama and Loebanr III makes up less than 3% of the total pottery assemblage, while at Bir-kot-ghundai red wares account for more than 20% of the assemblage (Stacul 1978, 140). Mat impressions on the base of pieces have been recorded from all the sites from PIV in Swat, and also at Ghaligai from pieces assigned to PIII.

Pits have been uncovered at all the sites except Ghaligai, and are dated to Period IV in the Swat chronology, with the exception of Aligrama, where the pits are assigned to PV. Interpretations of these pits are important, and have varied according to the excavator. Stacul assigns a dwelling function to the larger pits uncovered at the Swat sites with such comments as the "recent discovery at Loebanr III of various shapes of dwelling- and storage-pits" (1977, 251). He also suggests that such dwelling pits may have been used for various purposes, having noted that near "the fire-places, on the bottom of Pit H4 and Pit H7, have been…recovered two terracotta figurines representing the humped bull and a potsherd with a graffito of the same animal. On the basis of this evidence, we may suppose the performance of some cult ceremony around the fire-places" (1980b, 72).

When discussing Timargarha 3, Dani says that the mixed fill of the pits uncovered there meant that it "was not possible to take them as sacrificial pits" (1967, 113) but rather the contents led him to interpret them as refuse pits. At Balambat, Dani said that the presence of pit circles on the surface of the site 'obviously' suggested that they were dwellings (1978, 48-9). The sites, both in Dir and Swat, also have structures of stone walls in various phases. Even at Ghaligai some evidence for a dry stone walled building was uncovered outside the rock shelter itself (Stacul 1987, 33). At Bir-kot-ghundai, Kalako-deray and Loebanr III the stone walls are interpreted as a later occupation phase, succeeding the dwelling pits in PIV.

Quernstones, grindstones, mortars and other stone implements, such as the shouldered hoe from the early layers at Ghaligai, that could indicate either a manipulation of gathered plants, or even agricultural activity, are really only considered of significance at Kalako-deray, where the large numbers of the grindstones compelled the excavators to develop a new theory of economic organisation (Stacul 1994b, 238-9). These items will be considered with other artefacts that may have a bearing on economic activity at the sites, as part of the environmental data. This type of tangential information about plant exploitation is important because in situations where direct plant and animal material may be lacking, or assemblages small, alternative approaches to understanding the subsistence need to be included in the analysis (Reid & Young, 2000).

The nature of change between PIII and IV in the Swat chronology, and also between IV and V (Stacul 1987, 112), or at sites such as Kalako-deray, between IV and VII (Stacul 1995) has been linked to the migration and settlement of completely new populations. This includes theories of the spread of the Indo-Europeans in Baluchistan (Rana Ghundai and Dabar Kot), Afghanistan (Mundigak), Northern Iran and Soviet Turkmenistan (Stacul 1969, 56-7). Replacement of fine pottery with coarse ware in particular is seen as indicative of cultural change. Stacul also says that on the basis of the radiocarbon dates, the oldest Neolithic culture in Kashmir determined the arrival of the 'barbaric' phase (layers 17-16) at Ghaligai (1969). The material remains have also been considered by the excavators in terms of their similarities in cultural packages to areas such as Northern Iran and Kashmir, and this is discussed generally in terms of influence, or the diffusion of ideas.

The influence of Burzahom and the Kashmir Neolithic on the development of the Swat Neolithic is thought to be considerable (Stacul 1994a; 1987; 1977). The presence of black-grey ware vases with mat impressions on their base in Burzahom phase I and Ghaligai strata 16 and 17 apparently shows a geographic and chronological connection (Stacul 1969, 63). Stacul emphasises the connection between the

Neolithic of Kashmir and this period in the proto-history of Swat.

However, the dates so far assigned to Burzahom place the pit phase rather earlier than in Swat by nearly one thousand years (Allchin & Allchin 1982, 111). At Burzahom, the pits, referred to as pit-dwellings, have been dated using radiocarbon estimation at c. 2920 BC, and were associated with grindstones. Following this phase, a period of occupation characterised by the same stone industry, but with evidence for mud and mud brick houses has been uncovered. This phase of above-ground occupation structures has been estimated as continuing up to 1700 BC, which is the approximate beginning of the pit-dwelling phase in Swat according to Stacul (ibid., 113). In their summary of the Burzahom and Swat pits, Coningham and Sutherland (1998, 179) note that the Burzahom pits are dated to between c. 3000-1550 BC, and the large pits from Swat sites have been dated to between c. 1700-1400 BC.

Despite the difference in the dates, a strong link is found in the use of dwelling and storage pits at Burzahom, those at Loebanr III, and the presence of black-grey burnished ware, brown gritty ware, rippled rim, and basket impressed pottery types at both (Stacul 1977, 251). It is interesting that Stacul is suggesting that the influence of the Kashmir Neolithic is evident at Ghaligai, in PIII strata, and at Loebanr III in the later PIV. Given the time lapse between Kashmir and Swat, Stacul may now be overstating the direct links. According to Thapur, there are connections between Swat and Kashmir in the first half of the 2nd millennium BC, and in turn, also similarities to the late Neolithic of northern China (ibid., 251). These similarities include pit dwellings and blade and stone tools, and it is thought to be significant that there are jade beads from both regions, but only from Period IV (ibid., 251). Green jade has also been found at Ghaligai from Period IV (ibid., 252), but not from any later period at any of the sites. Stacul explains a move away from the Kashmir influence, and evidence of different affiliations as the "collected evidences suggest not only the spreading of a new stylistic horizon, but cultural and probably ethnical [sic] changes too" (ibid., 252). Links between NWFP, China and the northern Kashmir Neolithic are made on the basis of pottery similarities, and also the geographical connection.

Ghaligai 16 and 17 have been dated to after the end of the Indus Civilisation (Stacul 1969, 57), and there is a focus on decline within the Northern Valleys region after the end of the Harappan Period, and this is linked to wider social upheavals and migration activity. Stacul claims that there are also indicators that the differences between periods may be the result of differences in the socio-economic base of the groups to the fore at each different period. Stacul's explanation of this is that farmers and livestock controllers were prominent in the culture that took over on the plains - as opposed to the tribal groups of a pastoral, nomadic background that form the basic cultural groups in the mountains.

Copper items were recovered in small quantities from PIV layers at most of the sites in Swat, and the equivalent in Dir, and while iron was recovered from PIV layers, it was not significant until PV, where it was recovered along with copper, before dominating the metal finds recovered. Prior to PIV, Harappan style pottery is noted at Ghaligai, but there is no evidence for bronze technology. This apparently is because "it is quite probable that the human groups that produced the pottery of strata 19 and 18, made use of metal tools as well as the fast wheel, like the peoples of the Harappa culture, which we referred to for various comparisons involving this horizon" (Stacul 1969, 57).

4.8 Summary and conclusions

The similarities between the sites discussed above do indeed suggest that they are part of a cultural complex preceding and then associated with the Gandharan Grave Culture, that occurred not only in both Swat and Dir, but also in areas to the north and east (Allchin 1993; Khan 1993; Stacul 1966). PIV in the Swat chronology (c. 1700-1400 BC) is significant in this study for a number of reasons. It is a period of discontinuity with previous periods and with those following, explained largely by human movement, if not invasion. The relative homogeneity of much of the material remains has led to comparisons with sites such as Burzahom, suggesting an 'Inner Asian' connection, and what is in effect a closed region, confined to the lower ranges of the Hindu Kush and Himalayas (Stacul 1994, 1987). At the same time however, pottery types such as the rippled rim ware have been found in both PIV layers at the Northern Valley sites, and at the Bala Hisar of Charsadda. The changes in structural forms, from pits to above-ground walled rooms, and an expansion of the are occupied has been suggested as evidence of both an increase in, and greater permanence of population during this period.

Further, the recalibrated dates from Possehl (1994) given in Table 4.2 above, and those from the Bala Hisar of Charsadda (Chapter 5), allow for contact between the two regions, and these are included in the model developed in Chapter 9. By summarising the main features and artefacts from the Northern Valley sites, in particular from PIV, but also looking at their development and continuation, some understanding of the nature of archaeology in this region during this period has been reached. This is important because it underlies many of the models and theories about social and economic organisation and change in this area that have been developed, and is discussed in Chapter 9. These models and theories are based on an analysis of the structures, artefacts and environmental remains, but as will be demonstrated following a re-evaluation of the plant and animal assemblages, the use of new ethnographic data and the incorporation of new archaeological information and environmental data from the Bala Hisar can allow a very different predictive model to be proposed.

In addition to a summary chronology of the area, a summary of the pit data from Balambat (and Timargarha III), Aligrama, Bir-kot-ghundai, Kalako-deray and Loebanr III is presented in Table 4.3. The main purpose for exploring the pit data in some detail is that pits are significant in the interpretation of each site, except Ghaligai, where no pits have been recorded. The purpose and function of the numerous pits at each of the other sites has been the subject of a great deal of speculation and suggestion. Using factors such as size, shape, fill, lining and the relationship of pits to other features, their function, ranging from dwelling pits to refuse pits, and storage pits to kiln pits have been discussed. The function of the pits and their effect on the overall site interpretation are considered for each site in turn. This is then extended into a consideration of the importance of these interpretations in understanding the Swat and Dir Iron Age archaeology, and how these have been included in the extant models of social and economic organisation.

Table 4.3 Northern Valley Sites: Pit Data

Site	Period*	No. of Pits	Dwelling	Storage
Aligrama	V	11	no	yes
Bir-kot-ghundai	IV	6	yes	yes
Ghaligai	~	~	~	~
Kalako-deray	IV-VI	>32	yes	yes
Loebanr III	IV	24	yes	yes
Timargarha	IV-V (I-III)	20	yes	yes
Balambat	IV-VIII (I-III)	>18	?	yes

sources: Dani 1968, Stacul 1967-1997b
* period is Swat chronology, Timargarha chronology is given in ()
nb: function assigned by excavators

Chapter 5
Southern Plains archaeology

The main site that will be examined here is the Bala Hisar of Charsadda. There are a number of reasons for this choice, and these include the long archaeological sequence at Charsadda and the importance of the prehistoric sites here to the cultural development of the region (Wheeler 1962). The discussion of excavations at the Bala Hisar will here be confined to the excavation report by Wheeler following his 1958 excavation season (ibid.) and the work by the Universities of Bradford and Peshawar between 1995 and 1999 (Ali et al. 1998; Coningham pers. comm., 1999). These two excavations, some forty years apart, provide an opportunity to re-evaluate the interpretations not only of the archaeology of the Bala Hisar itself, but also the potential contacts between the Vale of Peshawar and the Northern Valleys, and the origins and development of urban sites in this region. The most recent excavations at the Bala Hisar have enabled collection of the new environmental material which is analysed in Chapter 7. This is important, because there are very few urban sites in this region that have had such analyses conducted on any scale.

The occupation sequence at the Bala Hisar was established initially by Wheeler (1962), and this will be discussed in greater detail below. While the excavations by the Bradford-Peshawar team have not greatly affected understanding of the internal sequence of the site, the use of radiocarbon dating has altered the accepted earliest occupation dates of the site (Ali et al. 1998, 15). Currently, it is suggested that the earliest dates relating to occupation at the Bala Hisar may be as early as the mid to the last quarter of the 2nd millennium BC (Ali et al. 1998, 7; Coningham pers. comm., 2000). Occupation continued through to the first quarter of the 1st millennium AD, and thereafter sporadically right up until the British occupation during the 18th century AD. The period discussed in this chapter is that from earliest occupation until the last quarter of the 1st millennium BC. This chronological section has been selected because it allows some overlap with the pre-Buddhist periods of the Northern Valley sites, up to PV of the Swat Chronology discussed in Chapter 4, Table 4.2. It also gives enough chronological depth to allow an examination of change over time within the material and environmental remains from the Bala Hisar itself.

Information about rural Iron Age sites in the Vale of Peshawar, or indeed in other parts of NWFP, is scarce. A recent survey of the Charsadda District (Ali 1994) concluded that while there are a number of sites from the Buddhist period, which is here dated to between the 1st and 5th centuries AD and later, there are no earlier sites (ibid. 94). Ali accounts for this by saying that the complete "absence of proto and prehistoric sites in survey is surprising. Perhaps it is due to the natural condition of the land, or it may be that the climatic or environmental conditions did not suit the folk of the Palaeolithic or Neolithic periods" (ibid., 93-4). However, two sherds recovered during this survey, with what are described as cut tops, may be classified as belonging to Wheeler's 'Rippled Rim' category (Ali et al. 1998, 15; Ali 1994, 67, 127, sherds 5.11 (EV-33)-NG/47 & 5.13 (EV-15)-JSA/80; Wheeler 1962, 39). Wheeler dated Rippled Rim ware to between c. 550 - 325 BC (ibid.). The recovery, therefore, of Rippled Rim pottery from these two sites suggests that rural sites were occupied at the same time as the Bala Hisar, and that the lack of evidence recovered during the recent survey may be due to other factors such as preservation.

Indeed, some 25 kilometres north of Peshawar the discovery and exploration of the protohistoric cemetery of Zarif Karuna has important implications for both the extent of the Gandharan Grave Culture, and also the nature of settlement in the Vale of Peshawar contemporary with the earliest occupation at the Bala Hisar (Khan 1993). Zarif Karuna is outside the Charsadda District, but close enough to the Bala Hisar to be of some importance in trying to understand the rural settlement in this area. The excavation and survey to date have revealed a cemetery with three distinct phases or periods of burial type, and these, along with the pottery recovered, have led the excavators to consider the site closely allied with the protohistoric graves of both Dir and Swat (ibid., 3-4, 25-54). To the east of Charsadda District, 27 kilometres east of the town of Mardan, the grave site of Adina was located and excavated. Although these graves are constructed from four enclosing stone slabs, they also have distinctive standing stones erected over them (Khan 1993, 161-2). They are not considered by the excavator to be a part of the Swat protohistoric type of cemetery, but rather are thought to resemble the megaliths that were reported from Dumlotti in the Deccan (ibid., 161). However, the strong resemblance in construction to other Gandharan Grave sites (Dani 1992, 407-9; Stacul 1987, 71), suggests that interpreting them as part this 'culture', known primarily through the excavation of the graveyards, is possible (Coningham pers. comm., 2000).

Following the discussion of the site of the Bala Hisar, the site of Hathial at Taxila will be briefly examined using similar analytical categories. The reasons for using this site are that it is broadly comparable to the Bala Hisar in terms of known occupation phases for the period under study, it is in a similar environmental setting, and there is published material available relating to the explorations and excavations here, and so provides data for comparison.

5.1 The Bala Hisar of Charsadda

The Bala Hisar of Charsadda is situated in the centre of a river plain, near the junction of the Swat and Kabul rivers, both of which have many nearby channels (Wheeler 1962, 3). The Bala Hisar is just one of a number of known mounds with structural evidence visible that occur in an area approximately two miles square in the vicinity of the modern town of Charsadda. Other mounds include Shaikhan Dheri and Shahr-I-Napursan (ibid., 1). The existing mound of the Bala Hisar is the largest of this group, rising over 20 metres above the surrounding plain. The site can be divided into two areas, first the higher western mound, or Bala Hisar, which has been interpreted as an artificial acropolis or citadel (Ali et al. 1998, 3). Second, across what Wheeler claimed was the old bed of the Sambhor River, is the lower eastern mound, which has been interpreted as a later suburb (Wheeler 1962, 23, 30). As noted in Chapter 3, the Vale of Peshawar is a semi-arid region, with no distinct season of maximum rainfall, and subject to extremes of temperature (Dichter 1967, 12-13).

Charsadda has been of interest to scholars for many years, since its identification by Cunningham as Pushkalavati or the Lotus City, one of the ancient capitals of Gandhara (Ali et. al. 1998, 2; Cunningham 1863, 89). This identification was made on the basis of historical geography, rather than excavation. It was known through the descriptions of Arrian and Calisthenes, the historians of Alexander the Great that in this region there was a great city. This city was one of the eastern capitals of the Persian or Achaemenid Empire, in the satrapy or province of Gandhara. The other eastern capitals have been identified as Taxila in the Punjab of what is modern Pakistan, and Kandahar in eastern Afghanistan (Helms 1982; Marshall 1951).

This identification of Pushkalavati was important for a number of reasons. Pushkalavati was the city that had been besieged for thirty days by the troops of Alexander the Great during his conquest of the Persian Empire. If the Bala Hisar was indeed Pushkalavati, then "Alexander's presence at the site and subsequent installation of a Greek garrison offer an opportunity to study the dynamics which were associated with the appearance of the Greeks in the region as well as offering an absolute historical framework" (Ali et al. 1998, 2). The association of this region with the Achaemenids and the Greeks led Wheeler to attribute the emergence of cities here to their presence. In effect, he said that the urbanisation of the Early Historic period, epitomised by Charsadda and Taxila, was not an indigenous development, but the result of Persian and Greek invasion and settlement (1963, 172; 1962, 5). This is important, and will be discussed in greater detail in the later section of this chapter dealing with the models of urbanisation that have been applied to this region, and how the archaeological data fit these models.

Cunningham initiated the first recorded exploration of the site in 1882, but it was not until 1903 that Sir John Marshall, Director General of Archaeology in India, excavated here (Marshall 1904). Marshall's excavations were confined to trenches on the summit of the mound and so only uncovered remains that were interpreted as Buddhist and Muslim (ibid.). It was not until Sir Mortimer Wheeler excavated here in 1958 that any attempt was made to understand the early occupation sequences located at the base of the mound. Wheeler's trench Ch.I, cut down from the top of the mound to the flood plain at the base and revealed 52 layers, eventually linking in to his trench Ch.III which was cut some 70 metres from the mound (Wheeler 1962, 18-22).

The most recent explorations at the Bala Hisar have been carried out by a joint team from the Universities of Bradford and Peshawar, and included two seasons of limited excavation. The aim of these excavations was to obtain chronological and stratigraphical information to clarify and update Wheeler's interpretations. These were described as "seven major themes arising from Wheeler's excavation at Ch.III. They were the course of the defensive ditch; the date of the defensive ditch; the relationship between the postern gate and defensive ditch; the nature of the ditch fill; the relation between the wall foundation trench; the nature of the ancient river channel flowing between the Bala Hisar and the eastern mound; and the search for evidence of later phases of walling or revetments around the site" (Ali et al. 1998, 6).

The importance of updating and refining the chronology of sites in NWFP and Pakistan prehistory is indisputable. The new ^{14}C dates for the occupation sequence at the Bala Hisar play a major role in re-interpreting not only of such issues as the indigenous development of urban forms during the Early Historic period, but also the contact between the Vale of Peshawar and the Northern Valleys of Swat and Dir. In turn, this allows new suggestions about the nature of subsistence and economic organisation, and the effect that an altered chronology has on interpretation will be expanded in Chapter 9.

However, the limited nature of the Bradford-Peshawar excavations at the Bala Hisar could also be seen as an area of concern. Allchin could have been aiming his criticism directly at these excavations when he said "With only a few exceptions, for example at Sonkh (Hartel 1976; 1993), early historic excavations in the second half of the twentieth century have been confined to cutting tiny sections through city ramparts or occupation deposits with a view to obtaining pottery sequences or chronological data" (1995, 7). The Bradford - Peshawar excavations achieved the stated project aims, allowing a re-interpretation of the chronology, and therefore a complete re-appraisal of the origins of urbanisation in this region, but they have done little to increase understanding of occupation of the site itself. In Chapter 2, the effect of the research design for this project on the environmental data was discussed, and not only has the scale and aim of the excavation affected the environmental data, but it has also necessarily limited any attempt to extend understanding of social organisation at the site.

In view of the nature of the 1994-6 excavations, the cultural sequence discussed here is based on the 1958 excavations. From his trench Ch.I Wheeler was able to establish a primary occupation and cultural sequence, which has been altered mainly in the lower layers and in details by the later work. The lowest, or earliest layers at the site are characterised by evidence of burning, that Wheeler called a 'conflagration' (1962, 20) and contained sherds of soapy red ware. Wheeler also says that during the earliest occupation phases, water was still flowing in the now dry Sambhor River bed, between the western and eastern mounds, and at this time the defensive ditch and associated structures were built (ibid. 18). The pottery associated with the earliest phase is found in layers up to 41, some 13 feet above natural, and Wheeler says that this represents two centuries, to which he assigns dates of ca. 530 -327 BC (ibid.). Mud brick walls and flooring are found in the layers up 43 and 42, then the structures uncovered are walls with pebble floorings (ibid., 20-21). One pit, X, falls within the period of interest here, and is cut into a mud brick floor in layer 27 then sealed by a renewal of the floor (ibid., 21).

Trench Ch.III uncovered what Wheeler interpreted as a defensive ditch, complete with postholes attributed to the presence of a timber lined postern and bridge (ibid., 27). Only the lowest layers of Ch.III were considered by Wheeler to be *in situ* deposit. Very little pottery was recovered from the ditch, and what there was comprised soapy red and rippled rim ware (ibid., 25, 28).

Exploration of what was thought to be the old course of the Sambhor River was the purpose of trench Ch.II, situated between the eastern and western mounds. Wheeler records that there was a deposit of some 13-14 feet from the present surface to the old river bed, which was shown by waterlogged grey sand (ibid., 23). The presence of typical early pottery in the lower layers, such as rippled rim in layer 15 and soapy red ware in layers 13 and 14, led Wheeler to suggest that filling of the ditch began in the latter part of the 4th century BC (ibid., 25).

Wheeler also conducted limited excavation on the lower eastern mound, after what he called modern "predatory cultivators" at Charsadda had uncovered the external wall of a mud-brick house (ibid., 28). Trenches Ch.IV and V revealed occupation evidence that Wheeler classified as later than the rippled rim ware phases, but he did note the presence of some soapy red ware in the lowest or earliest layers (ibid.). Five different phases were identified in the mud brick housing of Ch.V, and Wheeler tied these phases into the dating of the Bala Hisar on the basis of the pottery, and also to Bhir Mound at Taxila on the basis of architectural styles. The dates he assigned to Ch.IV and V were c. 300-150 BC for Phases II to V, with Phase I marginally earlier than 300 BC (ibid., 30, 32).

5.1.1 Wheeler's chronology

Wheeler used several key factors to develop his chronology at Charsadda. The siege of the city by Hephaistion and the troops of Alexander the Great allowed him to tie the site to an historical event, and so give at least one certain date, 327 BC (1962, 3, 34). However, this is predicated on the correct identification of Charsadda as Pushkalavati, one of the eastern capitals of the Persian Empire. This is Wheeler's one absolute date, and the other dates in his chronology are relative, based on the development of an internal pottery sequence, the presence of iron artefacts, and comparison of the material remains from Charsadda with those from sites such as Taxila.

The recovery of iron from the lower layers at the Bala Hisar allowed Wheeler to calculate that the city was founded during the 6th century BC (Wheeler 1962, 13). This in turn is derived from the conventional beliefs within South Asian archaeology about the date of introduction and the spread of iron in the sub-continent, which largely rest on the identification of iron in the Vedic literature (see Gordon 1960, 153). According to Wheeler "the Bala Hisar was in origin no designed acropolis but represents the gradual accumulation of floors and structures on a civic site in normal *tell* or *dheri* fashion" (1962, 12). Wheeler was also emphatic about the source of cities in this region, and this confirmed his suggested foundation date. He said that the available evidence strongly suggested that neither Taxila nor Charsadda itself existed prior to Persian rule in the region, in other words that cities in this period were a Persian import along with Persian rule (ibid., 5). Iron was also thought to have arrived with the Persians, and although iron had been recovered from Persia in contexts dated to the end of the 2nd millennium BC, Wheeler asserted that iron had not travelled without the invasion and conquest by the Persians (ibid., 33). Iron from the Bhir Mound at Taxila was dated by Marshall to c. 500 BC (ibid., 34; Marshall 1951, 538-9).

Wheeler was also able to recognise distinct pottery styles, and trace their introduction, peak and end of use within his archaeological layers. The main pottery types and these chronological changes are well expressed in his 'syncopated section' (Wheeler 1962, 36). This pottery sequence has been recognised as significant in the development of archaeology in this region, as many of the types recorded by Wheeler were new, and have since been noted in sites in the Northern Valleys (e.g. Ghaligai and Aligrama: Stacul 1969, 84; 1977, 175). In the absence of radiometric dates, Wheeler's chronology based on his pottery sequence has also been widely accepted.

The following are Wheeler's main categories of ceramics that fall within this period of study, with the dates and the position in the trenches that he reported. Rippled rim ware, where the rims recovered were thought to represent a fairly large, globular jar or cooking pot, made of brown buff ware, usually with a rough gritty surface and often wheel turned. The rims were sharply 'out-turned', generally an inch or more and were notched or rippled on the edge (ibid., 37). Rippled rim ware has been given dates of c. 550 - 325 BC (ibid., 39). Soapy red ware was the name given to sherds of a rich red ware with a distinctive soapy feel, and Wheeler dates it to c. 550 - 300 BC (ibid., 39). Wavyline bowls, so called on the basis of their grooved, wavyline decoration, appear in the early layers in soapy red ware, which allowed Wheeler to give them a date of c. 550 BC (ibid., 39). Northern black polished ware (NBP), or the local variation, was important for Wheeler to be able to tie the cultural sequence at Charsadda into contemporary developments in South Asia. NBP ware is fine, wheel made pottery, with a distinctive surface of grey or brown to black colour, with a polished finish, and was first noted in Early Iron Age sites in the Jamna-Ganges area. NBP sherds have been found as far north as Udegram in Swat, from 3rd century BC contexts. Approximately 18 were recovered from the Bhir Mound at Taxila, where initial interpretation has suggested that they occur at the beginning of the occupation c. 500 BC (ibid., 42).

Wheeler used the features and artefacts recovered during the excavation to support and develop his theory that the urban form of the second, or historic urban period in the north western part of South Asia was a direct result of Persian, rule and influence (1962, 34). As part of his culture-historical framework for interpretation he saw the similarities between Charsadda and Taxila as a direct result of their positions within the Persian satrapy of Gandhara. This model, and the effects of it on the interpretation of archaeology in this region will be discussed further in Chapter 9.

5.1.2 The Bradford - Peshawar chronology

The preliminary report of the first two field seasons by the joint Peshawar-Bradford team, which include two seasons of excavation, has been published (Ali et al. 1998), and the final monograph is in preparation. The aims and objectives of this project have also been discussed above, but they are worth reiterating because they do greatly shape their re-interpretation of the Bala Hisar and the development of urbanisation in the north west of South Asia following the collapse of the Indus civilisation.

Ali et al. (1998, 6) are explicit about their aims tying in with Wheeler's work and trying to either refine or test his findings. To this end they are therefore not testing new or alternative models of urban growth and development using data from the Bala Hisar, but aim to prove or refute Wheeler's model of urban growth and development as a result of Persian rule. The most important aspect of the Bradford - Peshawar investigation is the production of a series of ^{14}C dates, in particular from the base of Wheeler's defensive ditch. These dates, as discussed in the next section, allow a re-interpretation of Wheeler's model because they push the earliest dates for the site back considerably. This is extremely important because it allows the suggestion of an indigenous source for the second, or Iron Age, urbanisation in this region, and also important for this study, as it allows for contact between occupants of the Bala Hisar and the Northern Valley sites during Swat PIV, before the appearance of the Grave Culture.

The dates published by the Peshawar - Bradford team are as follows:

- 1270 - 930 BC at a 95.4% confidence interval (GRA-5246) on a bone sample from their trench Ch.VI (to the south of Wheeler's Ch.III), from a cut feature (61) east of Wheeler's defensive ditch (Ali et al. 1998, 7).

- 770 - 410 BC at a 95.4% confidence interval (GRA-5247) on a bone sample from the fill of the equivalent to Wheeler's defensive ditch (feature 55) from Ch.VI (ibid., 7).

- 770 - 370 BC at a 95.4% confidence interval (GRA-4219) from the fill of a post-hole relating to the equivalent of Wheeler's postern identified in trench Ch.VI. The sample material was not given (ibid., 10).

- 1260 - 900 BC at a 95.4% confidence interval (GRA-4210 (ibid., 14)) on a bone sample from the base of the cut for the foundation of the mud brick wall to the west of the defensive ditch (feature 174) in Ch.VI (ibid., 12).

- 80 - 220 AD at a 95.4% confidence interval (GrD-21831) from charcoal samples above a wall identified some 6m below the top of the mound (ibid., 14).

A number of criticisms of this dating sequence need to be addressed. First, are four dates (discounting the final date on the wall 6m below the top of the mound) sufficient to develop an alternative chronology? This is particularly important if the nature of the archaeological contexts from which the samples are being drawn is considered. The ditch fill in both Wheeler's trench Ch.III (1962) and Bradford - Peshawar trench Ch.VI (Ali et al. 1998, 10) has been disturbed, and further, the feature from which date GRA-5246 was derived was described as "part of what appears to have been the badly disturbed remains of a substantial cut feature, 61, at the eastern end of Ch.VI" (ibid., 7). Despite the assertion that the base of each feature (the ditch, 55, and the other cut, 61) were filled with 'in situ' deposits (ibid., 10), the use of material from a context such as a ditch fill for dating purposes could be questioned (Bowman 1990, 51-4).

The estimates resulting from the 1994-5 excavations and sampling now give a radiometric date for the ditch/es and associated postern and wall. These dates are earlier than those suggested by Wheeler on the basis of his relative dating sequence and comparisons with other sites in South Asia. They therefore refute Wheeler's model of the urban form as a Persian import and show that the Bala Hisar was occupied prior to Persian rule in this area. The nature of the ditches could suggest that the occupation was of a significantly concentrated and organised nature, as Callieri has for Bir-kot-ghundai (1992, 343). The site itself may have been occupied for a considerable (or alternatively a very brief) period before the construction of the ditch and other dated features.

Further to the published ^{14}C dates, there are at least four more dates from the 1996 excavation at the Bala Hisar, that have been presented at conferences, but are yet to be published. One of the aims of the 1996 field season was to learn more about the nature and date of the earliest occupation at the Bala Hisar (Coningham pers. comm., 2000) and this has been achieved. The excavators were able to divide the occupation sequence uncovered in trenches Ch. VIII and Ch. IX into three phases, and obtained radiometric dates from each phase.

These phases, their main features and dates are as follows:
- Phase A, the earliest phase comprising land surfaces with clear evidence for human occupation and activity. A calibrated date of 1420-1140 BC with a 95% confidence interval was calculated from a bone sample.
- Phase B yielded a calibrated date of 1380-1090 BC at a 95% confidence interval, again from a bone sample. In this phase, two circular pits, each with a depth and diameter of around 1 metre were uncovered. These pits were filled with river cobbles, and one interpretation is that they were foundation pits for a building.
- Phase C, the latest phase uncovered in these trenches has had two dates returned. The first, a calibrated date of 1260-990 BC, with a 95% confidence interval, was calculated from a sample from a small circular pit cut into the wall of an apsidal building. This building and associated features was overlain by a floor layer, on which an oven or furnace was built. The calibrated date from this later layer was 800-250 BC, also at a 95% confidence interval (Coningham pers. comm., 2000)

In terms of the earliest occupation at the Bala Hisar of Charsadda these dates are of great importance. While the ditches and defensive structures dated during the 1994-5 seasons had shown that the site was of significance prior to both the Achaemenid and Alexandrian eras, the dates from the 1996 season, which relate to occupation, show that the earliest dates estimated by Wheeler can be pushed back by around 1000 years (Table 5.1).

Given that the absolute dates do affect the sequence developed by Wheeler, how does this change the interpretation of the development and urban form of the Bala Hisar at Charsadda? Wheeler's interpretation was inextricably linked to his culture-historical theoretical base, within which invasion and diffusion are viewed as the prime influences of change. The importance of historical, or textual information as a means of linking archaeological evidence to known events or places was also crucial in shaping Wheeler's model. The identification of Charsadda with Pushkalavati by Cunningham was accepted and further supported by Wheeler (1962, 35) on what is quite tenuous evidence (Ali et al. 1998, 2). Yet having accepted this as historically attested fact, Wheeler then went on to tie the site into Greek and Roman accounts, and try to 'prove' the connection of the Bala Hisar with the city besieged by Hephaistion (1962, 35).

Wheeler was in fact guilty of shaping his working hypotheses and his interpretations to suit and support the theory that had already been formulated (Johnson 1999, 20-1, 38), and in doing so, was in keeping with the standard approach to archaeological interpretation of this period. The 'evidence' seemed to fit the theory very well - the recovery of iron artefacts from the lowest levels at the Bala Hisar, when he believed that the use and spread of iron within South Asia was connected with the Persian rule, allowed Wheeler to assign the date of the 6th century BC to the founding of the city (1962, 34). This tied in with both the conventional view about the spread of iron within the sub-continent, and with the founding of urban sites in this region being a result of Persian invasion.

In the absence of an explicitly stated hypothesis of urban development in this region, Ali et al. (1998) have explained their theoretical stance as one set on testing Wheeler's assertions. At its simplest, this means that they are proving or disproving the earliest settlement date of the Bala Hisar as during the 6th century BC. If the concerns over the archaeological contexts from which dating samples have been drawn are not considered major, then by pushing back the date range of the ditch fill to c. 1200 - 900 BC, and the occupation dates to c. 1420-1140 BC, they have clearly shown that Wheeler's model is incorrect. The ditch and associated features, and by implication some form of significant settlement, were established prior to Persian invasion in this region. This does allow an indigenous development of urban settlements in this region, a stance taken by at least one of the authors in discussion of the period in general (see Coningham 1995, 69-72).

The radiometric dates clearly show that Wheeler's earliest occupation date was too late by nearly a millennium (Table 5.1). Wheeler tied the earliest occupation phases uncovered on the tell itself in trench Ch.I with the lower contexts in the ditch (Ch.III) through a comparison of the pottery and other finds. In this way, he was able to assert the connection between the date of the ditch and of the occupation on the mound itself. This approach was also taken by the Bradford - Peshawar team in their analysis of the ditch fills, and so can be seen to demonstrate that the dates of the ditch fill can then extend the dates of the earliest occupation layers on the mound itself.

In terms of this work, the important consideration is that the radiometric dates do suggest the possibility that significant urban occupation at the Bala Hisar occurred considerably earlier than the previously widely accepted model of Persian introduction had suggested. Without further excavation however, an understanding of the extent or nature of the earliest settlement at the Bala Hisar is not possible. This is also true of Hathial at Taxila, and Allchin has said of both that it unclear "whether either constituted a 'city', or whether both are incipient cities, is something we can only speculate on at this time" (1995, 127).

Table 5.1 is important because it shows an overlap between the so far earliest known settlement in the Vale of Peshawar (Ali et al. 1998, 15; Vogelsang 1988, 109), and PIV of the

Table 5.1 Summary Chronology for the Bala Hisar of Charsadda & Taxila

Bala Hisar		Taxila	
Wheeler (relative dates)	Ali et al (^{14}C dates)	Marshall (relative dates)	Allchin (^{14}C dates)
Ch.X 6th C BC (foundation) Ch.III 4th C BC (defenses)	Ch.VIII Phase A 1420-1140 BC Phase B 1380-1090 BC Phase C 1260-990 BC 800-250 BC Ch.VI (= Ch.III) 770-410 BC	Bhir Mound 6th C BC Sirkap 2nd C BC	Sarai Khola 3100-2100 BC Hathial 2550-2288 BC up to 890-550 BC Bhir Mound end 5th C BC

sources: Ali et al 1998, Allchin 1993, Coningham pers. comm. 2000, Marshall 1951, Wheeler 1962

Swat Chronology. This overlap is shown on the basis of radiometric dates and confirmed by the presence of similar pottery types and other items, and will be further discussed and the significance explored in Chapter 9. This overlap, or link, is crucial in allowing the comparison of the environmental archaeological material recovered from sites in Swat and Dir and the Bala Hisar, which will be carried out in the following two chapters, 6 and 7. Not only does it permit the comparison of material from the different environmental settings, but it then makes the concept of contact, as shown in other artefactual remains a very real possibility.

5.2 Taxila

Taxila, situated in the Punjab rather than the Vale of Peshawar, NWFP, can be seen as a comparable site to the Bala Hisar for a number of reasons. These include the position of both on major trade routes within an agriculturally productive hinterland; the identification of both as eastern capitals of the Persian satrapy Gandhara; and the movement of the prime urban occupation to different sites within the same local region (Allchin 1993, 70; Marshall 1951, 1-3). Wheeler, in his analysis of the development of the Bala Hisar as an urban site was struck by what he saw as the great similarity with the development and form of the earlier periods at Taxila, saying "the story of Pushkalavati is essentially the story of Taxila" (Wheeler 1962, 15). It should also be remembered that the division of this area into the political and administrative provinces of NWFP and the Punjab are modern, and to some extent arbitrary.

Taxila has been the subject of more intensive and more extensive exploration and excavation than Charsadda. This is in part due its recognition by UNESCO and its status as a protected World Heritage site (Dani 1986, xi). Much more is also known about the sequence of regional development here, starting with the explorations at Sarai Khola, the earliest recorded site in the region which is situated south west of Taxila town itself. The sequence at Sarai Khola has been divided into three main phases, with PI, the Neolithic dating from 3360 - 3000 BC (all dates are calibrated radiocarbon dates); PII, the Early Harappan dating from 2909 - 2630 BC and PIII, the Late Kot Dijian dating from 2460 - 2090 BC (Allchin 1995, 125, 127). The dates of the Neolithic at Sarai Khola make it comparable with early Ghaligai in Swat and Burzahom in Kashmir (Allchin 1993, 70-1). Sarai Khola is also "distinguished by the presence of a number of dwelling pits" that have been assigned a ^{14}C date of 2700 - 2200 BC (ibid., 71).

At Hathial, Taxila, the earliest phase is equivalent to Sarai Khola PIII, the Late Kot Dijian, dated to 2550 - 2288 BC, and this is followed by what Allchin calls a 'hiatus' between the Late Kot Dijian and the next recognisable phase, that of the soapy red ware (1995, 127). Soapy red ware, or red burnished ware, has been recovered from layers at Hathial that have given ^{14}C dates of between 890-500 BC (Allchin 1993, 72).

At some point there was a move from Hathial to the Bhir Mound as the major focus of occupation at Taxila. Fussman (1993, 88-9) discusses the possible motives behind a move from the Bhir Mound to Sirkap, and these may also be applicable to a move from Hathial to the Bhir Mound. Hathial has also been interpreted as an acropolis for the Bhir Mound in the Indo-Greek period (ibid., 90). The available radiocarbon dates from the earliest occupation phases at the Bhir Mound indicate a date here of the end of the 5th century BC, but there is little evidence of any form to suggest Achaemenid influence at the site (Allchin 1995, 131). At the Bhir Mound, the oldest (and lowest stratum) is IV, and within this has been recovered Northern black polished ware (NBP), and along with further finds in stratum III of Gangetic single mould type terracottas, and terracotta ring wells better known in the Ganges region.

Marshall claimed that his excavations at the Bhir Mound confirmed that four successive cities were built on this site, before moving to Sirkap, during the period of control by the Bactrian Greeks (1951, 3). Marshall also said that according to the *Ramayana* (VIII, 101, vv. 10-16) Taxila "was founded at the same time as Pushkalavati in Gandhara by Bharata, ...

who installed his two sons as rulers in the two cities: Taksha in Takshsila and Pushkala in Pushkalavati" (ibid., 11). Based on the results of his excavations at Taxila, Marshall dated the earliest structures to around the 6th century BC, these being from the lowest level of occupation at the Bhir Mound. They are assigned to the Iron Age, and the nature of the stone work led Marshall to suggest a date preceding the 4th century BC.

While Marshall recognised the early status of the Bhir Mound, he did not recognise the significance of Hathial, and so was quite able to give an Achaemenid source for the development of the urban form at the Bhir Mound, and then attribute Sirkap, with a settlement date of the early 2nd century BC to the Indo-Greeks (Allchin 1993, 69). The dating of the earliest strata at the Bhir Mound to between the 5th and 6th centuries BC, also allowed the single iron piece recovered to be dated to this period (Marshall 1951, 538-9). As the number of iron objects recovered from the Bhir Mound did increase through the whole occupation span here, this also supports the introduction of iron with the Persians, and subsequent development. The 1967 excavations at Bhir Mound (Sharif 1969, 12) recovered iron from the earliest layers. Again, the presence of iron in what was interpreted as the original urban occupation phase was used to reinforce the circular argument of Persian origin for both the introduction of iron and the urban form (ibid., 13).

In terms of the urban nature of the Bhir Mound, in addition to its size, between 18-21 metres high, 1100 metres from north to south, and 670 metres from east to west, Marshall (1951, 3) also found evidence for street drainage and soak-away pits, that were interpreted as being part of a sewage treatment program. The architecture also included public spaces, or squares, and apparently public rubbish bins (ibid., 90-91). The Pillared Hall has been open to some debate about its function, with Marshall favouring a temple role. If so, he then claimed that this was the earliest discovered Hindu shrine at the time by several centuries. If this is accepted, then it poses some interesting implications for the spread of religion, and the co-existence of different types of worship. However, it has also been suggested that the Pillared Hall may have been the palace of an official or ruler, which could be seen as evidence for the hierarchical nature of the social organisation at the city. Some craft specialisation is also considered apparent in the concentration of the terracotta plaques showing male and female figures the Marshall calls deities (ibid., 98).

Marshall also described the city and street layout of the Bhir Mound as one that was 'haphazard' and 'irregular', and suggested that this was contrary to earlier interpretations attributing Taxila to Indo-Aryan influence (1951, 12). Marshall interprets the historical records, and indeed absence of evidence in these records as showing that "Taxila in the 4th century BC, although, no doubt, a large and densely populated city…had no architecture worthy of the name. So far as can be judged from the patches of remains uncovered, the lay-out of the city was haphazard and irregular, its streets crooked, its houses ill-planned and built of rough rubble masonry in mud, which, though neater and more compact than the masonry of the earlier settlement below it, was still relatively crude and primitive" (1951, 20). This haphazard nature of the town plan contrasts greatly with the geometric layout of the streets in the later city of Sirkap (1951, 89).

Allchin has developed a three stage model for the development of the urban form at Taxila. The first stage is one of growth from small settlements to something recognisable as urban during the 2nd quarter of the 1st millennium BC. This growth is attributed to the stimulus of the trade routes, and probably benefited from inclusion in the Achaemenid Empire. The transitional stage, from agricultural village, to city at the centre of a city state, is, Allchin suggests, represented by the soapy red ware, giving a date between c. 1000 - 500 BC (1993, 79). The second stage is one where the influence from the Ganges is recognised, and seen as very significant in the development of the Bhir Mound. The final stage is one of Indo-Greek influence, and is seen at Sirkap from the 2nd BC onwards (ibid., 79-80).

5.3 Summary and conclusions

The Bala Hisar of Charsadda is an important urban site, recognised as one of the major proto-historic and historic sites in the north western area of South Asia, and one of the major sites that has been assigned to the second urban period (Ali et al. 1998, 1; Allchin 1995, 16). The chronology of Wheeler placed the earliest occupation in the 6th century BC, but the more recent chronometric dates from the Bradford - Peshawar project suggest that an earlier occupation date of c. 1200-1400 BC is possible. If these earlier dates are accepted, then the lower occupation levels at the Bala Hisar are contemporary with PIV of the Swat chronology, and this is further supported by the pottery evidence where, for example, rippled rim pieces are found in both areas.

The excavation evidence from the Bhir Mound, and in line with more recent discoveries and interpretations, from Hathial at Taxila are largely in agreement with the Bala Hisar conclusions. At both sites the question of urban origins and the nature of early settlement remains. Are there likely to have been connections with the Late Harappan, as Shaffer (1993) suggests on the basis of his re-analysis of the recent archaeological data from these sites? Given the emphasis he places on cattle, and their role as either economic or cultural items, it may be possible from a careful analysis of the faunal and floral remains and decorated pottery from the Bala Hisar of Charsadda and the Northern Valley sites to make suggestions about how these sites fit into Shaffer's continuity model.

At what point can either site be considered urban? A comparison of the traits that have been used to establish the urban nature of archaeological sites is carried out in Chapter 9. Nevertheless without a great deal more data from the early occupation layers at the Bala Hisar, much remains in the realms of conjecture. The defenses have been dated to c. 1200-900 BC, and the presence of fortifications may be interpreted as showing that there is something there worth protecting.

However, an alternative interpretation could be that these represent defenses from flooding of the Sambhor River. The position of the site with regard to the rivers, and the fertility of the land might indicate a need to protect the settlement from rising waters rather than Alexander the Great's army. Yet it still does not answer the question of what was being protected. The surveys by the Bradford-Peshawar team, combining EDM and geophysical techniques, suggest that the defense may have followed the contours of the natural hill on which the site is positioned (Coningham pers. comm., 1999).

Similarly, Marshall's excavations at the Bhir Mound provided very little structural or other evidence that would define it as a major urban site (Childe 1950, 1957; Coningham 1995; McNairn 1980; Trigger 1972). There are no monumental buildings, with the possible exception of the palace/temple, which while large was not described as monumental. There is evidence for craft specialisation within that same temple, but if it is a palace, then it could equally be used to demonstrate the existence of a stratified society. There is no evidence for script, and the lack of regular grid planned streets may suggest a lack of scientific accomplishment. Yet from the other settlement evidence in the region, such as Sarai Khola, Pind Naushari and the other sites, it seems that the Bhir Mound did become a focus, and did grow into a recognisably urban form. Likewise, the Bala Hisar is set in a landscape that is rich with sites, albeit those identified to date have been assigned a largely historic date. The presence of Late Harappan pottery at sites in Swat may be a significant link in the development of the whole area. Late Harappan or Late Kot Dijian sites in the Taxila valley also suggest some kind of regional, if not national connection. Chapter 9 will present the arguments for the inclusion of Swat in a northern, or Inner Asian cultural development, particularly in relation to the earlier, Neolithic settlements. However, later evidence, both in Swat and in the southern Lowland urban sites suggests that contact with and influence from the south may have been just as important.

Even if the sites of the Bala Hisar and the Bhir Mound or Hathial do not fulfil all the traits suggested as necessary to define them as urban sites, they are nevertheless incipient urban sites, with very clear evidence for the later development of each as a major regional centre. It is hoped that by analysing new, and re-assessing old, archaeological environmental evidence, new interpretations of the nature of economic organisation will be possible. By comparing material from the Bala Hisar of Charsadda, which clearly did develop into a major urban site, with similar material from the sites in the Northern Valleys, which have been shown in Chapter 4 to range in size and status but not develop into cities, it is intended to demonstrate how faunal and floral remains can contribute to such interpretations.

Chapter 6
The Northern Valley sites: environmental material

This is a presentation and re-analysis of the environmental data from the Northern Valley sites of Balambat and Timargarha I in Dir, and Aligrama, Bir-kot-ghundai, Ghaligai, Kalako-deray, and Loebanr III in Swat. This has been carried out to understand the nature of subsistence in the Northern Valleys, and also to determine whether it is possible to characterise the environmental data from these sites as indicative of rural occupation sites. The results and conclusions can then be compared to the results and the conclusions from the urban site of the Bala Hisar of Charsadda. The environmental material from the Bala Hisar will be presented and re-analysed in the following chapter, and where comparisons are possible, these will be made in the discussion section of the next chapter.

The material discussed here forms the basis of the interpretations of subsistence and economy within the wider theories and models of occupation within these valleys. These theories and models will be discussed in Chapter 9, and it will then be demonstrated that the plant and animal remains, along with tangential archaeological subsistence material, can be interpreted in ways that substantially alter the accepted models. This is a crucial aim of this work - demonstrating that the re-analysis of this published data, and the comparison with new material from the Bala Hisar and the new ethnographic information, will allow new interpretations of the subsistence strategies, and through this the economic and social organisation of the occupants of both areas.

Table 6.1 summarises the type of material that was recovered from each of the sites, and indicates the method of recovery. Tables 6.2 and 6.3 show from which of the six sites from Dir and Swat plant and animal remains were recovered. This is intended to demonstrate presence and absence in each case, as numbers are not given here, but in the Tables at the beginning of Appendices 3 and 4 (plant and animal remains respectively).

That the summary data presented here in both graphical form and within the main text are reliant on the data that have been published by the excavators and specialists, does pose some problems in terms of standard information available for each site. For example, although the material from Ghaligai, Loebanr III and Bir-kot-ghundai has been published in one volume (Compagnoni 1987; Costantini 1987), comparable data for each of these sites is not always given. For example, age estimates on the basis of tooth eruption for pig are given for Bir-kot-ghundai, and sheep and goat age

Table 6.1 Summary of Environmental Material from the Northern Valley Sites

Site	Period	Faunal Remains	Hand Collected	Floral Remains	Flotation
Aligrama	IV-V^2	x	x	x	x
Bir-kot-ghundai	IV^2	x	x	x	x?
Ghaligai	I-III^2	x	x	x	-
Kalako-deray	IV^2	x	x	-	-
Loebanr III	IV^2	x	x	x	x
Balambat	III^1	-	-	-	-
Timargarha	II-III^1	x	x	-	-

[1] Swat chronology

[2] Timargarha chronology

Table 6.2 Summary of Plant Remains Identified from the Northern Valley Sites

Plant Taxon (Cereals)	Cropping Season	Aligrama*	Bir-kot-ghundai	Ghaligai*	Loebanr III
wheat (*Triticum* spp.)	W	Y	Y	Y	Y
barley (*Hordeum* spp.)	W	Y	Y	Y	Y
rice (*Oryza sativa*)	S	Y	Y		Y
oat (*Avena* spp.)	W		Y		
Legumes lentil (*Lens culinaris*)	W	Y			Y
pea (*Pisum arvense*)	W	Y			Y
bean (*Phaseolus* sp.)	S	Y			
Oilseed & Fibre linseed (*Linum usitatissimum*)	W				Y
Fruits hackberry (*Celtis australis*)	S?	Y		Y	
grape (*Vitis vinifera*)	S				Y

source: Costantini 1987, 1979
S summer sowing and autumn harvest (kharif crop)
W winter sowing and spring harvest (rabi crop)
* all periods

Table 6.3 Summary of Animal Remains Identified from the Northern Valley Sites

Animal Taxon	Aligrama*	Balambat & Timargarha	Bir-kot-ghundai	Ghaligai*	Kalako-deray	Loebanr III
Wild						
wild cat (*Felis* sp.)	Y					Y
tiger (*Panthera* sp.)						Y
barking deer (*Munticaus muntjak*)				Y	Y	Y
hog deer (*Axis porcinus*)				Y		Y
deer (*Cervus* sp.)	Y	Y			Y	Y
Himalayan goral (*Noemorhedus goral*)					Y	Y
markhor (*Capra falconeri*)						Y
hare (*Lepus* sp.)		Y				Y
porcupine (*Hystrix indica*)				Y		Y
snake sp.		Y				
Domestic						
dog (*Canis familiaris*)	Y		Y		Y	Y
pig (*Sus scrofa domesticus*)	Y		Y		Y	Y
zebu (*Bos indicus*)	Y		Y	Y	Y	Y
buffalo (*Bubalus bubalus*)	Y					
horse/donkey (*Equasasinus/Cabullus*)	Y	Y	Y	Y	Y	Y
goat/sheep (*Capra hircus/Ovis aries*)	Y	Y	Y	Y	Y	Y

sources: Bernhard 1967, Compagnoni 1987, 1979, Dani 1967, Jawad 1988
* all periods

estimates from epiphyseal fusion data for Loebanr III are provided, but this information has not been given for all species from all sites. This becomes even more difficult when work from other analysts is brought in, such as the Kalako-deray report (Jawad 1998), or the reports of both plant (Costantini 1979) and animal (Compagnoni 1979) remains from Aligrama, which are given in a different format, and only numbers animal species are given, and no quantification of the plant material has been presented. For this reason, as discussed in Chapter 2, analysis of the plant and animal assemblages will be confined to such estimates as MNI's (minimum number of individuals) and NISP's (number of identified specimens), which do allow for comparison between groups that may have had different criteria or analytical techniques applied (Bond 1994, 270; Hubbard & Clapham 1992, 119).

Each of the northern valley sites is looked at in turn, and any major trends or changes in the data noted. For each site, the interpretations of the environmental data given by the site excavators, and if appropriate the specialists, are be discussed. This is then followed by new interpretations of the data if possible. Following as examination of each site individually, they are looked at as a 'Northern Valley' regional group in terms of their environmental data.

Animal and plant remains from the Northern Valley sites (and indeed the urban site of the Bala Hisar) have by no means been collected systematically, as has been discussed in Chapter 2. This has a further effect on the comparability between each site in terms of period or date range being examined. It is important to be aware of for a number of reasons, it means that direct period to period comparisons of material from similar rural occupation sites in the Northern Valleys is not always possible, nor is it always possible to examine subsistence patterns and change at any given site in consecutive periods. However, and this is considered to be extremely important, an overall, generalised picture of subsistence strategies and food procurement in the two study regions can be developed. In any area where a particular study or sub - discipline is relatively new, generalisation is often the first approach, followed by greater data collection and thus greater detail, *once techniques and interpretations have been shown to be useful.*

As discussed in Chapter 1, archaeology in Pakistan is still greatly influenced by the legacy of Marshall and Wheeler, and still operates largely within the culture-historical paradigm. Environmental archaeology, as part of the new archaeology, gained increasing recognition in Europe and America in the late 1970's, and has since proved its worth. This is not the case in Pakistan, where art, architecture and cultural markers such as pottery, are all seen as worthy goals, by all but a minority. The work of this minority has been recognised as of great value, for example, the contribution of the archaeobotanical and archaeozoological analyses to the interpretation and significance of sites such as Nausharo and Mehrgarh (Jarrige 1997, 17)**,** but the fact remains that environmental assemblages from comparable periods from these different sites are not at this time available.

6.1 Ghaligai

With only 21 bones in total recovered and identified from Ghaligai, it is very difficult to establish any trends in these data beyond presence and absence. Nevertheless, the importance of the site itself in terms of material from more than one period, and the limited nature of the total database of bones in the Northern Valleys suggests that an examination of this limited assemblage is important. The bones recovered are from three periods: four from PI; six from PII and eleven from PIII. Two bones from deer species were recovered from PI, meaning that half the recovered bones from this period were from wild species and the other two bones were both identified as cattle. The single bone from PII was identified as belonging to a wild species, the porcupine. In PIII, all the bones were attributed to domestic species (ibid., 134). This may indicate a decline in the significance of hunting over time, and the accompanying increase in the importance of pastoral activity reinforced by the rise in the number of sheep and goat remains. Or it may indicate the butchery of hunted animals at the kill site, with no major bones being brought back to the occupation site. There were no sheep/goat bones from PI, a single bone in PII and four specimens in PIII.

Porcupine could have been either a source of food, porcupine hunting for food in Sindh is noted by Eates (1968, sited in Roberts 1977, 339), or they may have been killed as a crop pest (Compagnoni 1987, 136; Roberts 1977, 339). It is of course possible that both reasons contributed to the presence of porcupine bones in PII at Ghaligai, that an animal threatening crops might be trapped or hunted and then used as a source of food itself. Dog accounted for two of the total of eleven bones recovered from PIII. Dog bones at Bir-kot-ghundai were noted as having cut marks, and this has led Compagnoni to the conclusion that dogs were a source of food at this time (Compagnoni 1987, 137).

Cattle are the only animals for which specimens were recovered from all three periods at Ghaligai. This suggests that they were important here, probably both in terms of food, and also for traction. Stacul (1969, 50-1) suggested that Ghaligai may be a temporary or seasonal hunter-gathering site in the early occupation phases, and the presence of hackberry, a wild fruit, and bones from two species of deer, as well as the nature of occupation within the rock shelter itself support this. However, evidence for other structures outside the shelter, the presence of *Bos indicus*, or zebu, a domesticated animal and the recovery of both domesticated wheat and barley from PII and III respectively, strongly supports an interpretation that some form of crop and animal management was taking place in these later phases at least.

Sheep and goat were represented by bones in both PII and PIII, and in the latter, four bones were recovered, the same number as for cattle in this period. It is interesting that cattle are noted earlier than the sheep/goat, as cattle certainly require a greater maintenance input than sheep/goat (van der Veen et al. 1996, 252). However, the small assemblage precludes making any detailed interpretations of subsistence at Ghaligai, or about the way it may have changed over time.

One equid bone was recovered from PIII at Ghaligai, which may suggest that donkeys and horses were only of importance to the economy at a later stage in the development of the site (ibid., 134), which might be supported by the increased numbers recovered from two other sites during PIV. This could further be interpreted as part of a general move away from the hunting of large ungulates towards a more settled agricultural and pastoral economy. However, the presence of wild species at Loebanr III in PIV shows that they were still important, yet they were recovered alongside enough domestic material to show that domestic animals clearly dominated the economy.

The plant assemblage from Ghaligai suffers from similar sampling problems as the bone assemblage. Primarily, sampling was not systematic, as Costantini, the botanical specialist acknowledges (1987, 155), and overall, the assemblage is small. Samples were taken from three layers: layer 17, an ash deposit near a hearth in PIII (dated to c. 1900-1700 BC); layer 19, in PII (dated to c. 2500-2000 BC) and layer 21, in PIII (dated to c. 3000-2500 BC) (ibid.). Compagnoni describes a large quantity of barley straw and a single barley grain being recovered from the ash deposit of layer 17, PI. This is the only layer from which barley was recovered, and it may be suggested that barley has replaced wheat, which was recovered from the preceding PII from layer 19. Wheat grains, but no chaff or related material, were recovered. The two recovered weed seeds at Ghaligai, gromwell (*Lithospermum arvense*) and spurge (*Euphorbia helioscopia*) were also recovered from layer 19. The only recorded botanical material recovered from PIII were the stones of hackberry (*Celtis australis*). Of the latter, Compagnoni says "The quantity of stones and their concentration in heaps or deposits indicates that the fruits were gathered in natural growth areas and brought into the settlement to be eaten" (ibid. 156). While the concentration of stones does support an interpretation of their relative importance at this site during PIII, it does not necessarily mean that they were wild, and that the trees were not being manipulated in some way to enhance yield.

An examination of pottery from PI-III at Ghaligai by Costantini allowed the identification of rice grains, both as impressions in the clay and in the form of silicized remains (ibid., 162). Costantini interprets this as evidence for the presence of cultivated rice at Ghaligai, and so Swat, prior to PIV, when rice grains were recovered and identified from other sites. This early identification of rice, a summer crop generally agreed to require a degree of irrigation and agricultural planning, has implications for the nature of the economic organisation. Thompson (1996, 154) notes that wild rice typically grows in permanently wet areas, while domesticated rice is generally cultivated in areas of temporary flooding or inundation. The Northern Valleys correspond to the latter conditions, as the main rivers are prone to winter and summer floods. The presence of both summer and winter crops, and cattle (although not buffalo), may indicate more settled occupation, or even more regular seasonal use than Stacul has considered.

6.2 Aligrama

The animal bones from Aligrama have been published in a paper presenting them in conjunction with the Loebanr III bones (Compagnoni 1979, 697). Separation of the two assemblages has been on the basis of the MNI and NISP counts given in the Table attached to the paper (ibid., 701-2), and these figures form the basis of the analysis (given in Appendix 3). No other information, such as age at death estimates has been published for this material. The Aligrama material has been recovered from a number of contexts, the earliest from PIV right through to the 4th century AD, although for this study, the material from PIV, IV-V and V will be discussed (ibid., 697).

The total assemblage consists of 807 bones, of which 24 are from PIV; 93 from IV-V; and 690 from V, which shows a rather large increase towards the later period. However, the range of species represented by the material in the periods remains relatively constant. Two bones from wild animals were identified, one thought to be from a tiger in PIV and one from a deer species in PV (ibid., 697-8). The rest of the assemblage comprises domesticates, and in PV five bones identified as buffalo are recorded. This is very interesting as it indicates that buffalo were being utilised in this region, albeit in small numbers, and suggests that the conditions Patel (1997) and others describe as necessary for buffalo must have been met at this site. Also, because the separation between buffalo and cattle has been made here, in an early report, then the negative findings of buffalo at other sites are likely to be just that, and not mis-identification as cattle.

Cattle dominate the assemblage, both in terms of the NISP counts and the MNI counts, suggesting that cattle were of greater value here than sheep and goat. This is supported by the introduction of buffalo in PV, as a site that is successful in cattle husbandry may be prepared to commit the time and other resources to buffalo management. Pig and equids were also recovered from all the periods discussed at Aligrama, in roughly equal amounts that indicate that they while were present, they were not as significant as either cattle or sheep and goat. Dog, the only other domesticate identified, was not present in PIV, but is represented by a single bone from IV-V, and nine bones from V. It may be that this increase in dog, from an MNI of one to three in PV is linked to the overall increase in the size of the bone assemblage, and the presence of an estimated 38 cattle and three buffalo, as the dog may have been used to help control and protect the herd. As Dennell notes "the domestication of the dog perhaps affords us our earliest example of man reducing the effect of competitors by the use of biological control" (1983, 8). The presence of dog at both Loebanr III and Bir-kot-ghundai could also support this argument, in particular at the latter site, where the MNI estimate for dog in PIV is four, and the MNI for cattle exceeds that of sheep and goat (Compagnoni 1987, 134). Dogs are still important today in herd management, as shown by a number of those interviews recorded in Appendix 5.

With regard to the plant remains from Aligrama, the material was published in conjunction with the Loebanr III plant material, without quantification (Costantini 1979, 703). Sampling at Aligrama in 1972 and 1975 was described as 'trial sampling', and in the 1976 excavation, this sampling became systematic, although details have not been given (ibid., 705). Such lack of detail also makes it rather difficult to assign dates to the material, beyond being from the same periods as the faunal material, ie from PIV, IV-V and V, without distinction between them. There is one whole wheat grain, identified as *Triticum sphaerococcum,* 'numerous' naked and hulled barley, 'numerous' rice remains, lentils were 'abundant', and peas, beans and fragments of the hackberry were identified (ibid., 706-7). According to Costantini, the combined palaeobotanical evidence from Loebanr III and Aligrama, situated some four kilometres apart, "would suggest that the irrigated areas at the valley floor were exploited by means of differentiation of the crops in the two areas and of the gathering of wild plant at different altitudes during the different seasons of the year" (ibid., 708). The two areas Costantini refers to are first, the area around the river subject to flooding in winter and summer, and suitable for one rice crop per year; and second, higher ground subject to flooding only in summer, and so suitable for rice and wheat in summer and vegetables in winter (ibid., 707-8).

6.3 Bir-kot-ghundai

Like the assemblages from Loebanr III, Compagnoni says that the material is from occupation layers ascribed to Period IV of the Ghaligai sequence (1987, 131). It is unfortunate that little has been done to examine each site separately, to try to determine change over time, or even to describe any intra-site spatial variation in the animal remains. All the bone material discussed from Loebanr III, Bir-kot-ghundai and Ghaligai, are considered indicative of food remains on the basis of the (unspecified and unquantified) evidence of cut marks.

The total number of bones recovered and examined from Bir-kot-ghundai is 2047, the largest assemblage of any of the sites discussed. No bones or fragments identified as belonging to wild species were recovered from this site. The greatest number of bones 1464, or 71.5% were identified as cattle (again considered belonging to the single *Bos indicus* or zebu species). This in turn gave an MNI estimate of 46, giving an MNI percentage of 43.3, which places cattle as the single most important animal type at Bir-kot-ghundai, both in terms of NISP count and MNI (ibid., 134, 142).

Equid bones are of far greater importance at Bir-kot-ghundai than any other site examined. 158 bones were recovered, giving an MNI of eleven, nine of which were attributed to donkey and two to horse, with a possible identification of mule (ibid., 138). The presence of cut marks on equid bones has led Compagnoni to suggest a food use for these animals, but this is qualified by the animals being used for food once they are no longer fit for traction or transport (ibid., 131, 140).

Sheep and goat were represented by 300 bones recovered from Bir-kot-ghundai, which gave an MNI of 32. The bones account for 15% of the total bones, and the MNI for 30% of the total number of calculated animals (ibid., 134, 142). Of the mandibles examined, 25% from Bir-kot-ghundai had milk teeth, compared with 14% from Loebanr III. Compagnoni interprets the differences in age at death as one reflecting economic, if not social structure at each site "As with the pig, this proves that at Bir-kot-ghundai, sheep and goats were slaughtered at a much earlier age, on an average, than at Loebanr 3" (ibid., 145).

Compagnoni has assigned all the pig remains from both Loebanr III and Bir-kot-ghundai to domestic species (ibid., 140). At Bir-kot-ghundai, pig accounts for 5% of the total bones, numbering 105 fragments, and these were calculated to give an MNI of 13, which places pig as the third most important animal (ibid., 134). The analysis of the tooth wear from pig mandibles showed that here, some 77% of the specimens recovered were one year or younger at the time of death, and that more than 92% were under two years of age at the time of death (ibid., 140). Compagnoni uses the age at death of the animals, and the comparison of this between Loebanr III and Bir-kot-ghundai, to say that this "would imply, although one must proceed with caution in making such conclusions, that the inhabitants at Bir-kot-ghwandai … had more refined eating habits than their fellow men at Loebanr 3" (ibid., 141).

In terms of body parts recovered, the horse/donkey and cattle remains from Bir-kot-ghundai are clearly dominated by teeth and distal limbs, whereas the goat/sheep assemblage, although again dominated by teeth, shows far less disparity between the distal and proximal limb bones recorded (ibid., 136, 142). Pig also show slightly more proximal limb bones, but this is not as extreme as the difference in either cattle of horse/donkey (ibid., 140). This may suggest that while sheep/goat are being used for meat, hence the recovery of the upper limbs or body parts with greater meat on them, the use of cattle primarily for secondary products may affect which skeletal parts enter the archaeological record (ibid., 140, 142). Other explanations could include the use of cattle long bones for artefact manufacture, or breaking up for marrow extraction.

Soil samples intended for the recovery and analysis of plant remains were collected at Bir-kot-ghundai, and these are all attributed to PIV of the Swat Chronology (Costantini 1987, 156). Therefore, the botanical material from this site has been treated as a single assemblage, with no differentiation made for layer, thus no discussion of temporal change in plant based subsistence is possible, or even spatial distribution across the site. What is possible is a discussion of the range of species present, and how this compares overall to the other sites with plant remains (Aligrama, Loebanr III and Ghaligai) in the Northern Valleys, and with the Bala Hisar of Charsadda in the Vale of Peshawar.

Twelve wheat grains from PIV, 37 grains of barley, and 39 grains of rice were recovered (ibid., 156-157). Interestingly, Costantini does not mention the recovery or identification of any weed seeds, wild plant remains or chaff. As these categories have all been noted at Ghaligai, it is presumably correct to suggest that they were not present at Bir-kot-ghundai. In turn this suggests that the PIV plant remains are likely to be the result of cereals processed else where, and so the grain itself was the main object, or survivor here. In terms of seasonality, both wheat and barley are considered winter crops, and rice a summer crop. Although the small numbers in this sample make assessing the relative importance of each crop difficult, the numbers of rice and barley grains recovered and identified, clearly outnumber any other taxa. The only other recorded plant remains from Bir-kot-ghundai are a single oat grain and five lentils from PV (c. 1400-1100 BC) (ibid., 157). When considered with the plant material from PIV, this shows that the occupants of Bir-kot-ghundai were exploiting at least three cereals and one legume crop.

6.4 Kalako-deray

To date only faunal remains have been collected, identified, analysed and published from the site of Kalako-deray. Stacul mentions in his excavation report for the 1989-1991 fieldwork that wheat and barley grains were recovered from pit B7, and identified by Costantini (1993, 75). However, this appears to be the only consideration of plant remains from the site to date. All the faunal material has been recovered from five pits (B5; B7; B17; C8; C24), and these have been assigned to Period IV in the Swat chronology based on the pottery typology and a radiocarbon date (Jawad 1998, 265). This dating is important as it gives a firm base from which to compare other environmental assemblages from the northern valleys, and to compare the assemblages from the northern valleys and urban site from the Vale of Peshawar, the Bala Hisar of Charsadda, as discussed above.

A total of 131 faunal bone fragments were recovered (and identified) from the pits at Kalako-deray, and they are reported as bearing cut marks made by sharp (ie iron) tools (ibid., 265-6). Various reasons are put forward by Jawad to account for the small total number of bones recovered, that relate to depositional rather than taphonomic or recovery factors. The animals represented by the bones recovered from these pits are cattle (including both *Bos* and *Bubalus* species, although Jawad does not provide separate estimates for these two, but combines them under one heading of *Bos* sp.), sheep and goat (grouped together for analytical purposes), deer including barking deer (*Muntiacus muntjak*) and the Himalayan goral (*Naemorhedus goral*), pig and bird. As can be seen from the analyses in Appendix 3, cattle dominate the faunal assemblage, accounting for 55.5% of the total identified fragments, with sheep and goat accounting for 38.2% of the identified assemblage (ibid., 268).

It is important to note that while Jawad has not given any breakdown of the cattle category, he does include water buffalo in this section. This could be interpreted as showing that both were present in the pits at Kalako-deray, or that no attempt was made to identify them, or that the identification could not be made. The presence of water buffalo is important in terms of interpretation issues to be discussed below, but it should also be remembered that the skeletal differences between cattle (*Bos* spp.) and buffalo (*Bubalus* sp.) are often difficult to determine on archaeological material (Higham 1975; Thomas 1986, 129).

As Jawad says, the low NISP count of 53 for cattle at Kalako-deray does not permit any extensive statistical analysis or comparisons to be made, but the measurements taken from the available material can be compared to that from the other Swat sites (1998, 271-2). Using phalange measurements, it appears that the Kalako-deray cattle were smaller than those at either Loebanr III or Bir-kot-ghundai (Appendix 3). Jawad also reports that there were no juveniles among the cattle, and those that could be aged were classified as mature animals, over seven months and up to 42 months (ibid., 272-3). However, on the basis of an MNI estimate of five Jawad says that cattle were the main source of meat food at Kalako-deray, reaching this conclusion through a comparison of the MNI estimates of the identified species (ibid., 271). Using the results of both the MNI estimate and the age at death estimates, Jawad suggests that during PIV at Kalako-deray "cattle was most probably primarily kept for secondary purposes, i.e. milk and traction etc., and were killed for food at an advanced age stage when they became less functional" (ibid., 273). However, the greater numbers of cattle bone compared to other species leads him to conclude that they were the preferred meat for consumption (ibid.).

From the sheep and goat remains, an MNI of nine was calculated (ibid., 274). Using a data set of 14 mandibles, Jawad was able to calculate that all but two animals were in the adult age stage (based on Payne 1973, tooth wear), with one classified as an infant, and one classified as belonging to the old age stage (Jawad 1998, 274). Analysis of the epiphyseal fusion of the long bones tended to confirm the tooth wear conclusions (ibid.). From this information, while acknowledging the lack of data relating to sexing of specimens, Jawad suggests that male sheep/goat were kept primarily for meat consumption, while females were kept primarily for breeding and dairy purposes (ibid., 276-7).

The four deer specimens have been attributed to at least two different species: Barking deer and the Himalayan grey goral, as well as two indeterminate deer antlers (ibid., 279-281). The presence of the deer, and also pig suggests that hunting was an aspect of food procurement, at Kalako-deray in PIV, although not dominant (ibid., 277). Jawad does not indicate whether the pig specimens are from domestic or wild animals, but as an MNI of one is calculated from the remains, they are certainly not as important in terms of subsistence as either cattle or sheep/goat.

Jawad's conclusions are necessarily limited by the small amount of faunal material recovered, however, it is possible to begin to recognise some trends in the data. While cattle bones dominate in terms of the numbers identified, at Kalako-deray, the difference between the identified numbers of cattle and sheep/goat bones is the narrowest of all the sites where figures are available. The MNI estimates, despite Jawad's claim to the contrary (ibid., 271), suggest sheep and goat were of greater importance in terms of meat here, as the small size of the cattle restricts their contribution in terms of amount of meat per animal. That both the sheep/goat and cattle groups age at death estimates were dominated by adult specimens is interesting. This may mean that neither group were of primary significance for meat, and that secondary products such as milk and wool or traction were of greater importance. Alternative explanations could include the absence of young animals from the site due to seasonal occupation only, or the disposal of bones from young animals in areas of the site other than pits. The small assemblage also makes determining trends in the body parts recovered difficult, and as likely to be the result of taphonomic and recovery processes as depositional influences.

No report on plant remains from Kalako-deray has been published, however, as noted in Chapter 4, a significant number of querns, and other stone artefacts that have been interpreted as associated with plant processing have been recorded here (Stacul 1994a, 235). It has been suggested by the excavator that one of the possible functions of Kalako-deray could have been as a manufacturing centre for stone tools (ibid., 239), although the evidence for occupation was also noted. That local schist is available at Kalako-deray is also pointed out, however, it seems that no analysis of possible sites receiving the products has been made to date.

6.5 Loebanr III

Of the 1180 bones examined at Loebanr III, cattle account for nearly 61% of identified material, while sheep and goat are the second most numerous group, accounting for 34% of the identified bones (Compagnoni 1987, 132-133). Table 6.5 and Appendix 3 give greater details of the NISP and MNI counts for the animal bone assemblage from Loebanr III. When the MNI for each of three main domestic animals is calculated, it appears that sheep and goat, with an MNI of 48, were present in greater numbers than either cattle or pig (ibid., 142-3). Compagnoni stresses the importance of sheep and goat at Loebanr III, especially when compared to the results from Bir-kot-ghundai, and says that this dominance may be indicative of differences in social organisation at the two sites (ibid., 133). When considering both the total number of identified bones and the MNI estimate, Compagnoni says that "these data show that shepherding was, at the time, one of the major economic activities of the people, especially at Loebanr 3, even though the contribution made by sheep and goats to the diet, given their smaller size, must have been secondary to that made by cattle" (ibid. 145). This, of course, rests on the assumption that both sheep/goat and cattle were kept primarily for their contribution to the diet.

Teeth account for the greatest number of identified skeletal elements recovered for sheep and goat, and this is followed by proximal limb bones. While taphonomic factors may account for the high numbers of teeth, it may be possible to suggest that decisions about meat may have influenced in part the presence of the proximal limb bones. The abundance of cranial remains from sheep and goat allowed a calculation of age at death on the basis of tooth wear, and from this Compagnoni has suggested that the 14% of mandibles with milk teeth, and the fusion data from long bones, show that sheep and goat were being killed at a later age than those at Bir-kot-ghundai (ibid., 145).

Compagnoni explains the classification of all the cattle bones to *Bos indicus* as "the referring of the archaeological cattle to *Bos indicus* L. is due to the discovery at Bir-kot-ghundai of a bifurcated spinous process of thoracic vertebra, held by the authors to be a characteristic, though not exclusive, of the humped cattle, a deduction supported furthermore by the numerous figurines representing the zebu cattle which were recovered both at Loebanr III and at Bir-kot-ghwandai" (ibid., 142). None of the large bovid material has been identified as buffalo. As mentioned above in relation to the identification of cattle at Kalako-deray, distinguishing between zebu and buffalo is not always straightforward, and this could greatly affect interpretation of the assemblages, and in particular conclusions about social organisation (Meadow pers. comm., 1999).

With regard to the cattle skeletal elements which were recovered and identified, distal limbs were most numerous, then proximal limbs and teeth. There was less disparity between the element groups for cattle, than for sheep at Loebanr III. The age at death for cattle at Loebanr III is calculated on the basis of tooth wear, where one out of fourteen mandibles only has deciduous premolars, and with very few unfused postcranial bones recovered (ibid., 142).

This has led Compagnoni to suggest that the majority of cattle were living to adulthood before slaughter, which could support an interpretation of the importance of cattle primarily for secondary products.

The pig remains at Loebanr III are dominated (three-quarters of the identified material) by skull and tooth fragments. From the tooth eruption recorded, and the unfused nature of the surviving long bones, Compagnoni has suggested that most animals were slaughtered while very young (ibid., 140). However, in general the tooth eruption from the domesticated animals at Loebanr III, when compared to that from Bir-kot-ghundai, indicates that the animals were older at death at Loebanr III, and this has led Compagnoni to suggest that eating patterns were different at the two sites (ibid., 141). Equid bones were also recovered from Loebanr III, in the form of one adult horse tooth, and four other maxillary fragments that have been attributed to a young adult Equid species (ibid., 134, 137-8).

The range of wild species present at Loebanr III is great, especially in comparison to any of the other Northern Valleys sites, but as Compagnoni points out, each species is represented by a very low number of identifiable fragments, with only deer and porcupine having an MNI estimate of more than one (ibid., 134). This has led Compagnoni to conclude that while hunting obviously occurred, it was not significant when compared to domesticates. The very low NISP of each wild species may be the result of butchery occurring largely off-site for wild animals, or the possible use of long bones and horns of the larger wild animals such as deer, for other purposes (ibid., 134).

As Loebanr III has the greatest number and range of animal remains, so it has yielded the greatest number and range of plant remains. The material recovered and identified from Loebanr III includes cereals, legumes, fruits and weed seeds (Costantini 1987, 165), and these are detailed in Table 6.3 and Appendix 4. All the material analysed has been recovered from layers within pits at the site, and all these layers have been dated to PIV (ibid., 158). The plant assemblage discussed here comprises charred seeds only.

Of the cereal grains identified, barley was the most numerous type present, accounting for 66% of all identified cereal, and 43% of the total seed assemblage (ibid., 159). Wheat, identified by Costantini as *Triticum sphaerococcum*, and rice were also recovered. It is interesting, given Weber's earlier analysis of the occurrence of rice in relation to other cereals on South Asian sites, where he found that rice "usually occurred in the form of isolated finds and rarely in sites where large number and varieties of species were found" (1991, 26) that it has been recovered from contexts with other grain types. Both lentil and pea were also recovered, suggesting that legumes were important in the plant base of the diet of the occupants of Loebanr III. Costantini tries to draw an even more significant conclusion from their presence, saying that the "finds from Loebanr 3 form an important chronological and geographical reference point for the diffusion of this species in the Indo-Pakistan sub-continent where, according to de Candolle (1883:pp 257-8), *Lens culinaris* arrived following the Aryan people" (1987, 159).

The recovery and identification of linseed seeds is important not only because these are the first to be noted in this region, and linseed can be used for both textile and oil production, but also because of the size of the seeds recovered. Costantini suggests that the measurements obtained from these seeds show that they are bigger than uncultivated linseed seeds, and further, that size increases not only with cultivation, but also with irrigation of the crop (ibid., 160). Two grape seeds, or pips, have been identified, and Costantini attributes them to *Vinus vinifera*, or cultivated grape (ibid., 160). Cultivation of grape is thought to be part of what Costantini describes as a "widely developed agricultural system" in this region in the 2nd millennium BC (ibid.).

With regard to the cropping seasons, both winter or rabi crops (wheat, barley, lentil, pea and linseed) and summer or kharif crops (rice and grape) have been recovered. Thomas (1986, 1999) and Reddy (1997) suggest that the presence of different seasonal crops may, depending on local growing conditions, indicate a seasonal or permanent occupation at a site. This is discussed further below, where material from other rural sites is considered. At Loebanr III, the presence of a number of specimens from both main crop seasons could be interpreted as evidence for year round sedentary occupation. Loebanr III certainly has a wide range of winter and summer crops, as shown in Table 6.3, although as discussed in Chapter 8, modern transhumant groups from both Swat and Dir indicate that seasonal movement does not preclude the potential for both summer and winter crops to be grown in these valleys.

Of the weed seeds recovered and identified, Costantini says that little of significance can be learnt about either crop or environmental conditions (1987, 160). The weeds are very similar to modern weed seeds recovered from the site, indicating that major environmental or agricultural changes are unlikely to have occurred.

6.6 Timargarha 1 and Balambat

With regard to analysis of environmental material, Balambat presents a number of major problems. While Dani noted that animal bones had been recovered from all of the Timargarha and Balambat trenches "we have not been able to find specimens of corns in our excavations nor have we been able to get a full report on the animal bones" (1967, 26). The only animal bones that have been identified were recovered from graves. While animal bones and charcoal were noted as part of the fill of many of the pits at Balambat occupation site, these have not been identified or analysed further. In addition to the animal bones from the graves, there are references to structures, pottery forms and other artefacts that may be interpreted as relating to subsistence activities. In the absence of direct evidence, this tangential material and the interpretations that have been given by the excavators, will be discussed.

The graves that contained animal bones are all from the burial site of Timargarha 1, and the species identified were horse, goat, sheep, stag, hare and snake (Bernhard 1967, 332, 370). Of the five graves that contained animal bone specifically identified, four were fractional burials, and thus assigned by Dani to the latest Gandharan Grave period, PIII, dated to between 1000-500 BC (Dani 1992, 397; 1967, 71). The remaining burial contained both evidence for a cremation burial and a later fractional burial, so could belong to either PII (1400-1000 BC) or PIII in the Timargarha chronology (ibid.). There is some variation both in the species in the graves, their ages and the skeletal parts recovered, and these are summarised in Table 6.4.

The analysis of the human remains in these graves suggests that they were all adult at the time of death, but the fragmentary nature of much of the material made sex determination difficult, if not impossible (Bernhard 1967, 299-300, 303, 305, 308, 313-4). Dani has suggested that the snake bones may be of either ritual significance, or the result of a snake entering the grave by accident and then dying there (1968, 32, 35). It is also noted that while no animal terracotta figurines were recovered from any of the graves, figurines in the form of humped bulls, ram's heads and what are described as cats, have been found in occupation layers at Balambat (Dani 1967, 31). Given the sampling and recovery problems from the excavation of the occupation site, it is not possible to try to argue any great significance from this.

Table 6.4 Animal Remains from Graves: Timargarha 1

Grave	Species	Element	Age Estimate	Sex
109	goat	skull and mandible fragments	6-8 months	female
		distal limb (?) fragments		
125	horse	'extremity' bones		
	stag	mandible	ca. 8-10 years	male
	sheep/goat	two teeth, ruminant		
137	sheep	long bones		male
	sheep/stag	tow humerus fragments, metacarpus		
	hare	mandible and limb bone fragments		
183	snake	tail vertebrae		
149	sheep	skull fragments, mandible	ca. 10 years	
		proximal and distal 'extremities'		

source: Bernhard 1967, 370

In his summary of the structures at Balambat, Dani notes that there are rectangular and circular storage rooms, and this is taken as evidence for grain storage, although there appears to be little support for this interpretation (1967, 26). The presence of certain pottery types familiar to the excavators was taken as evidence of subsistence as "food habits can be guessed from the flat dish (*thali*), which is very convenient for eating rice" (ibid., 28). The location of the site, and other Gandharan Grave sites, is also seen as significant in terms of the subsistence of inhabitants, as the settlement areas are all found on the slopes of hills, from the lower foothills to the rivers. As the graves themselves are concentrated in a relatively small area, and the occupation evidence stretches from the hills to the river, this suggests, according to Dani, "that the river water must have been used for limited irrigation, as is the practice in the present day" (ibid., 26). This is extended further when Dani suggests that the use of the lower land for cultivation would have necessitated the use of higher slopes and the tops of the hills for grazing cattle (ibid.).

Among the artefacts from Balambat are two items of iron, described respectively as a 'gardening tool' and a 'sheep shearer'. Both were recovered from Period IV contexts, thus making them later than the main grave phases, and so after 500 BC according to the Timargarha chronology (Rahman 1968, 275). The terracotta animal figurines from Balambat are dated slightly earlier, to PIII according to context, and although again it is difficult to try to extrapolate much in the way of information from these finds, it may be possible to suggest that clearly both the exploitation of plants and animals were known to the occupants of both sites.

The recovery of animal bones from graves at Timargarha, and the presence of the animal terracottas and other items at Balambat do show that a range of species, both domestic and wild were known to the occupants of these sites. Inferences about the plant based diet are harder to make on the evidence, but given the position of the site on the fertile river plain, the possibility of cultivation, and even irrigation is certainly there. It is difficult to move beyond this point on the basis of the evidence available. Indeed, while snake is thus far unique to Dir in terms of animal bone assemblages, the rest of the material does not contradict any of the findings from the other sites in Swat discussed here. This is important, because it allows the suggestion that these sites, and their occupants were part of a loosely connected group, the dynamics of which will be discussed further in Chapter 9.

6.7 Environmental material and interpretations from rural Harappan and Bannu Basin sites

The following is a brief summary of some of the environmental analyses and their subsequent interpretations from some of the recent work that has been carried out on rural sites in South Asia. The two main areas where environmental research has been conducted are in the Bannu Basin, where survey and excavation was initiated by the British Archaeological Mission to Pakistan and the University of Peshawar (Allchin et al. 1986), and continued by the Bannu Archaeological Project (Thomas 1999); and the ongoing exploration and recording of Indus Valley Civilisation or Harappan sites (eg Saraswat 1986; Weber 1999). This is not intended as an in-depth investigation of environmental conditions or subsistence strategies from either of these types of sites. Rather, it is intended to place the material from the Northern Valleys in context, both in terms of how typical the recovered material is within the wider geographical area of South Asia and when compared with material from some other rural prehistoric sites. While it is clear that from a chronological aspect the sites are not comparable (all the Bannu and Harappan sites are earlier than the Swat or Dir sites) they are all considered rural settlements by the excavators, and given the smallness of the Northern Valley samples, it is very important to bring in supporting data in the attempt to try to determine whether it is possible to characterise these sites 'rural' on the basis of their environmental assemblages.

The extensive sampling and recovery programme at the Harappan site of Rojdi in Gujarat has provided a great deal of botanical material, from which Weber has been able to build up a picture of plant use, and agricultural change (1999, 1991). Rojdi is described as a village within a farming community, dated to the mature Harappan, c. 2500 - 1800 BC (Weber 1991, 37). On the basis of the analysis of the plant remains which, while they are dominated by summer or kharif crops such as millets, also have a proportion of winter crops present, Weber says that Rojdi is likely to be a permanent settlement rather than a seasonal site (1991, 51). Both the faunal and the floral remains recovered from Rojdi show a far greater diversity than at any of the Swat or Dir sites. In addition to the domestic staples of cattle, sheep/goat and pig, a range of wild mammals, reptiles, birds and fish, and over 60 types of plants were identified (1991, 47, 62-3). Due to the stratigraphic resolution of the samples, Weber has been able to argue that there was a significant shift in plant based strategies over time, and these were tied into socio-economic factors such as an increase in the number of settlements, but a decrease in their size leading to an alteration in the supply and demand of agricultural products (1999, 821, 824).

At two further Harappan sites in Gujarat, Oriyo Timbo and Babar Kot, Reddy (1997, 181-4) has been able to demonstrate that differences in the botanical record can be interpreted as evidence of different occupation patterns at each site. From Oriyo Timbo, the archaeobotanical assemblage is dominated by weeds and summer crops such as millet. At Babar Kot there is evidence for both summer and winter crops, as well as crop processing by-products. This has led Reddy to suggest that Oriyo Timbo was a seasonal, pastoral site, consuming but not cultivating, while Babar Kot was a permanent, sedentary site where both cultivation and consumption took place (1997, 184).

In terms of dominant species at these and other Gujarat sites, cattle are the most significant, both in terms of absolute bone counts and estimated MNI's (Thomas et al. 1997, 767; Weber

1991, 47), with sheep and goat also important, and pig the third most important species. Cattle is thought to be of such importance both because they are well suited to conditions in the area covered by the Indus Civilisation, particularly zebu, (Meadow 1984, 37), and because they provide a range of primary and secondary products. The high proportions of adult to juvenile cattle have been interpreted as an indication of the importance of secondary products such as milk and traction (Thomas et al. 1997, 770).

Of the Bannu Basin sites that have been recorded, those falling into the Early Harappan or Kot Dijian period (c. 2500 -2040 BC) are most relevant to this study, and this includes sites such as Islam Chowki, Tarakai Qila and Lewan (Thomas 1999, 312). The analysis of environmental material from these sites shows that while barley dominates the botanical assemblage, wheat was also present, and from Tarakai Qila, lentils were also recovered (Khan et al. 1991, 10; Thomas 1999, 315). The plant remains are thought by Thomas to show the clear seasonality of agriculture in this region, being almost all winter crops, that would be sown in autumn or winter and harvested in late spring (1999, 315). In a discussion of current environmental conditions in the Bannu Basin, it is pointed out that "High evaporation rates in summer mean that, without irrigation, agriculture is virtually impossible" (Thomas & Knox 1994, 95). Further, it is suggested that the conditions that might prohibit the growing of summer crops here, are also likely to prevent pastoral activity (ibid.).

The faunal material indicates that fewer wild species were being exploited in the Kot Dijian period than earlier. Again cattle dominated the assemblages from almost all the sites examined, and water buffalo is present in the Kot Dijian assemblages (Khan et al. 1991, 26; Thomas 1999, 314). Only at Tarakai Qila were sheep/goat numbers equal to the cattle numbers (Thomas 1999, 314). Interestingly, at Tarakai Qila around 30% of the sheep/goat bones were from juveniles, suggesting their slaughter for meat, while both cattle and buffalo were represented by adult specimens suggesting their use for secondary products (ibid.).

The site of Lewan in the Bannu Basin has been interpreted as a factory of stone artefacts, on the basis of the quantity and type of lithic material recovered both through surface collection and excavation (Allchin 1986, 63). While a number of querns and grindstones were recovered, "there is little evidence from these excavations to suggest that Lewan was the focus of an economy based upon crop production, although plant foods grown elsewhere were probably used at the site" (Thomas 1986, 123). With regard to the faunal assemblage here, cattle account for some 81% of identified bone, with sheep/goat next in terms of numbers, and then equid remains, and also buffalo. Indeed, while pig were noted as significant in the assemblages from the Harappan sites, bones identified as equid are significant in the Bannu Basin sites (ibid., 133; Khan et al. 1991, 9). In the analysis of the Lewan faunal assemblage, Thomas (1986, 134) noted the bovine body parts were dominated by distal limbs, whereas the caprine remains are dominated by both ribs and proximal limbs. However, it is pointed out that as the majority of the sample was recovered from what have been interpreted as rubbish pits on the site, then there are many biases in the material (ibid., 133). It has also been suggested on the basis of the environmental and economic analysis and interpretation that Lewan may combine both seasonal and sedentary settlement (Thomas 1986, 135), with the presence of buffalo as indication of at least an element of permanent occupation.

6.8 Summary and conclusions

In terms of overall numbers, cattle followed by sheep and goat, and barley, lentils and rice are the most significant of the plant and animal remains recovered from the Northern Valley sites. It is possible to make some very general statements about subsistence, namely that domesticated animals are far more important than wild, although deer have been recovered from all the sites except Bir-kot-ghundai; cereal crops comprise the most significant category of plant foods, with both summer and winter crops present. Wheat, barley and rice have been recovered from three of the four sites where plant material has been analysed. This suggests that both pastoral and agricultural elements were important, and that year round settlement may be part of the subsistence approach. However, the presence of both winter and summer crops, combined with a pastoral element may also be achieved through seasonal strategies, or a combination of seasonal and sedentary strategies. Compagnoni interprets the presence and numbers of both sheep/goat and cattle as the result of their primary importance as food (1987, 142, 145), although the relative meat obtained from each has led him to suggest that cattle were the more important in terms of contributing to diet. It is of course possible to suggest that secondary products were at least as important.

However, within this region, the individual sites show at least as much difference as similarity. At Bir-kot-ghundai, only cereals grains have been recovered, with no chaff or other processing material evident. No wild animals were identified, and the age at death of cattle, sheep/goat and pig could suggest a higher status site, eating only young animals. Loebanr III has the greatest range of both plants and animals, and the number of identified bones from wild species suggest that hunting was certainly of some significance, if not necessarily economic importance, here. The wild animals recovered from the sites can be almost all found within the Northern Valleys today, which suggests a local origin (Compagnoni 1979, 699-700).

The older age at death indicated for cattle, sheep/goat and pig could be, as Compagnoni suggests, indicative of less refined eating habits (1987, 141), or of a site where secondary products were of greater importance. The presence of both domesticated grape and rice, summer crops, and linseed, a winter crop, all requiring a degree of labour intensity, suggests that some form of agricultural planning and organisation would have been necessary for this year round exploitation of crops.

Modern irrigation in the Swat and Dir is mainly by either terracing and small channels linked to the river system, or by rainwater. The importance of water management in the occupation of marginal environments has been recognised and discussed in archaeological studies (Kennedy 1995, 275). Irrigation and water management have also been shown to play a significant role in urban development and population expansion. For example Morrison's (1995, 157; 1993, 147) work at the Medieval City of Vijayanagara in southern India utilised the analysis of a range of environmental and structural material in relation to irrigation. By identifying changes in the pollen sequence and charcoal intensities from periods during which structural remains associated with water management such as reservoirs and sluices were constructed, Morrison was able to demonstrate the link between the rapid growth of the city and water control, which permitted an intensification of agriculture. Systematic archaeological study of the antiquity of terracing and channels has not been carried out to date in Swat or Dir, but archaeological and ethnographic investigation in the Libyan Desert have shown that irrigation using small stone walls and floodwater farming techniques has been successfully practiced for several millennia (Gilbertson & Hunt 1996, 191).

The lack of quantification of either plant or animal remains precludes detailed analysis of the material from Aligrama, but again a range of both summer and winter crops were recovered, along with both wild and domestic animal species. Domestic animals dominate the assemblage, and this is the only site from which buffalo has been separately identified (from PV material). Cattle are estimated as having a greater MNI count than sheep/goat during all the periods, and this is the only site where cattle exceed sheep/goat in this analysis. The total number of identified bones from each period at Aligrama increase over time, with nearly five times more recovered from PV than PIV, which may indicate an increase in population, or possibly an increase in permanent population. Unfortunately, the lack of chronological detail for the plant assemblage means determining changes in summer or winter crops over time is not possible.

Kalako-deray has a very limited range of animal material, with the majority belonging to domestic taxa. The wild species all belong to deer, which suggests hunting for either food or sport. Cattle dominate the bone count, and sheep/goat the MNI estimate. However, the count of four deer bones translates to an MNI of four, while the cattle bones count of 61 identified fragments translates to an MNI of five. This type of imbalance calls into question the true meaning and importance of MNI estimates.

No reptile, fish or shell remains have been identified or analysed from any of the Swat sites, however the presence of snake at Balambat and one bird fragment from Kalako-deray shows that animals other than large wild or domesticated species were known, and exploited, though not necessarily for food, or in any significant quantity. The influence of sampling and recovery strategies must be considered here.

Costantini's interpretation of the three Swat sites of Ghaligai, Bir-kot-ghundai and Loebanr III in terms of the botanical remains, suggests they represent different stages in the local agricultural development, as the Ghaligai samples are from PI-III, Bir-kot-ghundai and Loebanr III are from PIV (1987, 161). This is largely the result of the biased sampling strategy where samples were taken only from those contexts considered archaeologically interesting (ibid.). Costantini is firm that the material from these sites represents established agriculture. Despite the recovery of only fruit stones from the earliest period investigated, and only fruit stones and a few wheat and weed seeds from PII, it is asserted that wheat is unlikely to have been the only crop known, and that the barley recovered from PIII "should be interpreted as a continuity of its presence rather than a new event" (1987, 161). Hackberry stones, although not recovered following PII from Ghaligai, are known from later periods at Aligrama. The presence of rice from PIV onwards, is interpreted as an important crop used to make use of land that was not suitable for other crops, and so allowing two crops or cropping seasons per year (ibid., 161).

When comparisons are made with rural sites from other areas, namely the Bannu Basin sites and the Harappan Gujarat sites briefly discussed above, further interesting results appear. At Aligrama, the slightly greater numbers of horse/donkey remains than pig, place it as the third most important animal at the site, which is more like the assemblages from Bannu than the other Swat sites. In contrast at Loebanr III, the numbers show clearly that cattle, sheep/goat and pig are important, while horse/donkey numbers are the same as those of the other wild and background animals. Cattle dominate in both analyses carried out on the bone assemblage, which tends to resemble the Harappan sites. Kalako-deray has the least disparity between cattle and sheep/goat, and no horse/donkey remains at all were identified. The large numbers of querns and other stone tools that could be associated with plant processing from Kalako-deray may allow an interpretation of the site as a centre of stone tool manufacture. The absence of plant remains is puzzling, yet this is also the situation at Lewan in the Bannu Basin which was interpreted by the excavators as a centre of stone tool manufacture. Again, at Lewan, as discussed above, cattle dominated the animal bone assemblage, but equid remains were also recovered.

It seems clear that no single pattern of subsistence can be discerned in the environmental remains. Rather, each site has animal and plant assemblages that are unique and, beyond some gross trends, such as the importance of cattle, sheep/goat and cereals, there are as many differences as similarities. This suggests that attempting to predict a rural package of subsistence is not possible, as the responses to even similar environmental conditions are very different. This will be developed further in Chapter 8, when the modern ethnographic interviews are presented and analysed, and then both the archaeological and the modern material will be combined in Chapter 9, the discussion and development of a model to explore subsistence in the study area as a whole.

Chapter 7
The Bala Hisar of Charsadda: new environmental material

The analysis of the plant and animal remains from the Bala Hisar, and interpretation both within and between regional contexts is essential for an increased understanding of subsistence strategies from the two environmentally and geographically contrasting regions of NWFP. Also important to this research is the comparison of material from the Bala Hisar of Charsadda, described as a major urban site of the Early Historic period (Ali et al. 1998, 1; Allchin & Allchin 1997, 249-50), with that from the Northern Valley sites, such as Bir-kot-ghundai which in later periods is described as urban (Callieri 1992, 343), or rural agricultural settlements, such as Kalako-deray (Stacul 1994b, 238). This comparison is important for determining whether urban and rural sites can be distinguished on the basis of their environmental assemblages.

The environmental material from the Bala Hisar of Charsadda is summarised here, and full details are given in Appendices 1 and 2. This data is analysed to see if there are any apparent trends in terms of taxa and their relative proportions, and changes in these over time. This is primarily a discussion of material from a single site, as there are no other comparable sites known or excavated in the Vale of Peshawar from this period. While there is no published environmental material from comparable excavated periods at Taxila, Marshall did record tangential evidence for agriculture, mainly in the form of querns and grinding stones and socketed hoes from the Bhir Mound (1951, 485, 559). Therefore, as part of the discussion and assessment of the Bala Hisar material to determine whether it can be considered urban in nature, the environmental material and conclusions from some urban Harappan sites will also be briefly considered.

Undertaking such comparisons with Harappan material has a number of drawbacks, as discussed in Chapter 6. Firstly, the periods under consideration are different. The Harappan period extending from c. 2500 BC to 1900 BC, while it is now estimated on the basis of the most recent chronometric dates that the earliest settlement at the Bala Hisar was c. 1400 BC (Coningham, pers. comm., 2000). Further, many more Harappan urban sites than Early Historic period urban sites have been excavated, therefore more is known about the economic organisation and the subsistence base of these sites. However, in the absence of more directly comparable material, consideration of urban Harappan analyses and interpretations does provide some indication of the types of plant and animal bone assemblages that are being recovered from pre-historic urban sites. There have been a number of suggestions about the political and social nature of Harappan organisation (eg Kenoyer 1998; Piggott 1950; Shaffer & Lichtenstein 1995), which will undoubtedly affect the nature of subsistence practised. While the nature of urbanisation in the Early Historic period, as known from work at Taxila and Charsadda, is somewhat different, some of the basic definitions of 'urban' within archaeology (covered in Chapter 9) are common in the two periods, and gives at least an initial starting point for comparing the subsistence base of each.

While Wheeler makes no reference to any environmental material in his 1962 report of explorations at the Bala Hisar or the eastern mound at Charsadda, the Bradford-Peshawar team, during their excavations between 1994 and 1996 collected both animal bones and plant remains. The sampling and collection strategies have been discussed above (Chapter 2), and this necessarily affects the assemblages. However, as been iterated in previous chapters, it is better to work with what is available, and try to demonstrate that such material can give rise to valuable interpretations and analyses, than to ignore what has been recovered. As noted in Chapter 2, and at other points in this volume, the presence and absence of species (and where possible, changes in these over time) is the primary aim of analysis in relation to the existing data sets.

When archaeological integrity is considered, only a very few contexts containing animal bones, and even fewer containing plant remains, can be analysed here. This is because many of the later contexts from all trenches at the Bala Hisar have been exposed to mixing and robbing. This problem has been noted by both the Bradford - Peshawar team (1998, 7-8) and

Table 7.1 Context Descriptions for *in situ* Archaeological Deposits from which Plant and Animal Remains were Recovered from the Bala Hisar of Charsadda

Context	Phase	Date	Description	Faunal Remains	Floral Remains
1071	A	1380-1090 BC	fill of pit (cobbles)	x	
1077	A		layer	x	
1065	B	1260-990 BC	burnt layer		x
1063	B	1270-930 BC	layer of burnt earth	x	x
1056	B		layer	x	
1076	B		layer	x	
57	B		fill of ditch	x	
60	B		basal fill of ditch	x	
71	B		degraded mud brick	x	
1031	C	800-200/	layer		x
1032	C	770-410 BC	area of burnt material		x
1043	C		fill of pit		x
1052	C		levelling/floor foundation	x	x
72	C		ditch fill	x	
73	C		ditch fill	x	
79	C		ditch fill	x	
80	C		basal ditch fill	x	

sources: Ali et al 1998, Coningham pers. comm. 2000

Wheeler (1962, 39). Table 7.1 summarises the contexts from which plant and animal bones have been recovered.

These contexts represent a range of occupation layers including pit and ditch fills, some evidence for burning, and structural layers, such as the degraded mud brick wall (context 71) and levelling layer (context 1052). One of the difficulties already noted with the Bradford - Peshawar excavations at Charsadda was the size of the trenches, which often precluded understanding and interpreting the nature of some of the material uncovered (Coningham pers. comm., 1999).

7.1 The plant material

Plant remains were recovered from contexts in both major trenches by the Bradford - Peshawar project. However, the total assemblage from contexts that represent *in situ* archaeology from phases relevant to this research period is very small. Table 7.2 is a summary of the identified material from contexts appropriate to the aims of this chapter. Rice grains and husks have been noted in some quantity from later contexts, however, only fragments of husks have been recovered and identified from one *in situ* archaeological context, and this means that conclusive statements about rice agriculture or trade are not possible. The presence of rice even in such small quantities does indicate the potential for rice at Charsadda, whether it is being imported to the city from sites in the Northern Valleys and processed prior to use at the Bala Hisar of Charsadda, or whether both grown and processed here.

The contexts from which plant remains have been recovered are significant primarily because of their connection with the dating samples. This is important because it allows the link between subsistence trends at the Bala Hisar in the Southern Plains to those in the Northern Valleys during PIV. While the nature of the contexts, as layers and ditch fills are not ideal for the analysis of plant material, they do allow an assessment of what was present at this site.

The presence of rice (*Oryza* cf. *sativa* L.) (Thompson 1996, 164-183), lentil (*Lens culinaris* Medik) (Renfrew 1973, 113-115; Zohary & Hopf 1988, 85-92), wheat (*Triticum* spp.) (Zohary & Hopf 1988, 16-28) and weed seeds of dock (*Rumex* spp. L.) (Schoch et al. 1988, 142-3), goosefoot or fat hen (*Chenopodium* spp. L.) (ibid., 50-53), bedstraw (*Galium* spp. L.) (ibid., 178), at the Bala Hisar, albeit in such small quantities, gives a range of crops similar to the range recovered from the Northern Valley sites. However, overall there is less variety of plant foods identified from the Bala Hisar than from the Swat sites, where barley, pea, and bean were noted, and also non-food crops such as linseed. The types of crops being grown and consumed, indicated by the plant macro-remains assemblage suggests a dominance of winter, or rabi crops, with the exception of rice. There appears to be little in the way of change in the plant assemblage over time, until the appearance of the rice husk in a context from Phase C, which is dated to between c. 800-200 BC. Otherwise wheat and lentil are constant, and present in all phases from which plant material has been recovered and identified.

The presence of wheat at the Bala Hisar, and the absence of barley may be linked to the urban nature of the economy. Weber (1999, 823) notes that wheat "was most common at Harappa between 2600 and 2000 BC, during the period of integration. This was when the city was most densely populated and when agriculture and herding activities may have been occurring further from the site". Further, barley is recognised as a more hardy crop, which is more suited to extreme environmental conditions than wheat (Thomas 1999,

Table 7.2 Plant Material from the Bala Hisar of Charsadda

Phase*	^{14}C Date	Context	Identified Material	Cropping Season
A	1380-1090 BC	1063	lentil (*Lens culinaris*)	W
A		1065	wheat (*Triticum* spp.)	W
C	800-200/ 770-410 BC	1031	wheat weeds	W
C		1032	lentil	W
			rice (husk frag) (*Oryza sativa*)	S
			weeds	
C		1043	wheat	W
C		1052	wheat	W
			lentil	W

S summer sowing and autumn harvest (kharif crop)
W winter sowing and spring harvest (rabi crop)
* phasing not related to Swat or Timargarha phasing or chronology

315, 317). Weber also points out that due to some of its physical characteristics, such as having softer straw than wheat, and the need to mill or grind it prior to de-husking, makes barley (in contrast to wheat) more suited to animal fodder than human food (1999, 823).

7.1.1 Urban Harappan plant material

Weber (1999, 823) has developed a model from the analysis of the archaeobotanical remains from urban Harappan sites, which suggests that at urban sites with high, dense populations, agricultural activity is carried out further from the city itself, thus cereals may be brought from distant villages, or areas of agriculture. This distance may be a factor in the relatively narrow range of cereal types recovered, and also likely to be responsible for the presence of human food crops, rather than animal fodder. It is possible to see similarities between the plant assemblage from the Bala Hisar of Charsadda and the urban Harappan material and models.

The dominance of wheat and the lack of barley at the Bala Hisar compares with an urban, non-pastoral economy believed to characterise at least some of the mature Harappan urban sites. Further, the fairly narrow range of cereal and other plant types from the Bala Hisar is in keeping with the urban Harappan model. However, due to the limited nature of the assemblage, both in terms of overall size and spatial and chronological coverage, it is necessary to remember that such comparisons should remain general in nature, and not be considered as a predictive model for either the Early Historic urban period, or the comparative Iron Age rural material.

7.2 The animal bone material

Animal bones were recovered from more relevant contexts than the plant remains, nevertheless they do not constitute a large assemblage themselves. In total, 57 bones were identified to species and element, and a further 248 bones were classified according to size of animal, and identified in terms type of bone (ie long bone shaft, articulation and so forth). Other bone material remained unidentified. Within the assemblage of identifiable bones from contexts that represent *in situ* archaeology, no fish, reptile or bird bones were identified, nor any bones from small mammals.

There are a number of possible reasons for the assemblage comprising material only from large mammals, and these include a basic collection and retrieval bias resulting from the way the material was collected (see Chapter 2 for a discussion of this) which operates in favour of big bones from big animals being recognised and collected. This bias can also be seen to be reflected in the body parts that dominate this assemblage, and this will be discussed in greater detail below. It could also be argued that at a large urban site, it is less likely that small mammals will be entering the archaeological record as a result of being, say crop pests (eg porcupines, see Chapter 6), as there is likely to be a greater separation between fields and food production and consumption. This may also account for the concentration of certain parts of the skeleton being represented (from all species present) while others are scarce, or absent. If slaughter, and at least some butchery is taking place away from the more densely occupied areas, then the representation of body parts will reflect this (Reitz & Wing 1999, 112, 202-4).

It is possible to distinguish some trends in terms of the species present, age at death and the main body parts represented, even within such a small assemblage. It is also possible to suggest some changes over time, although because the bones identified do come from a rather small number of contexts, and are small in terms of total number, such suggestions remain rather general. The material has been collected from two trenches (VI and VIII) at the Bala Hisar located at the base of the mound and adjacent to it. These trenches were approximately four metres apart, and subsequent interpretation by the excavation director, Dr Robin Coningham, has assigned the same phases to each. Therefore, the animal bone material has been grouped according to phase, and will be discussed primarily in terms of these phases. Table 7.1 gives each context and its interpretation for both trenches, from which animal bone has been recovered. It also presents the date ranges for the phases from which identifiable animal bones have been recovered from contexts with *in situ* archaeology, and these show that if the new and re-calibrated dates, discussed in both Chapters 4 and 5 and for the Northern Valley sites presented in Table 4.2, are considered, then the lower levels at the Bala Hisar of Charsadda can be seen to overlap with Swat PIV.

In terms of the total animal bone assemblage from the contexts with *in situ* archaeology, cattle clearly dominate among the 57 bones that could be identified to both element and species (see Table 7.3), comprising 47% of the assemblage. Sheep/goat make up 30%, and this is followed by buffalo at 14%, and then pig, 5% and deer, 4%. When these total counts are broken down into phases, cattle are still the most significant animal both in terms of the NISP and MNI calculations, with the exception of Phase C (the latest Phase under consideration here), from which no cattle bones were identified. However, a quantity of bone which could not absolutely be identified as cattle was recovered, and has been assigned to the 'large bovid' category of the bone classified according to size, discussed below. With such a small assemblage as this, some of the problems of MNI counts as a means of assessing the relative importance of different animals are clear. For example, while there is only a single deer bone from Trench VI, phase IV and there are seven buffalo bones, they both give an MNI estimate of one.

While Compagnoni (1987, 142) has assigned all the cattle bones recovered from the Northern Valley sites to zebu (*Bos indicus*) largely on the basis of measurements, there is no conclusive evidence (ie the presence of a bifurcated spinous process on the thoracic vertebra) given for this (see Chapter 6). It is possible that both zebu and non-humped cattle (*Bos taurus*), are present, and this may also account for some of the bi-modal size distribution in the cattle bones noted by Compagnoni (ibid.), which has been attributed to sexual dimorphism. Modern records and ethnographic evidence support the presence of zebu in common use around the Bala Hisar. A single articulated skeleton from the site of Gor Khuttree in modern Peshawar has also been identified as belonging to *Bos indicus* on the basis of bifurcated spinous processes of the thoracic vertebra (Young 1998, 143), which indicates the presence of zebu in urban archaeological contexts in this region, albeit from a later period. However, as no bifurcated spinous processes were recovered or identified from the Bala Hisar material, it cannot be asserted definitively that the cattle were zebu, or zebu alone. The usefulness of zebu in an agricultural landscape for ploughing and transport is demonstrated by the ethnographic interview results, which will be considered along with the archaeological data.

The relative importance of sheep/goat to cattle at Charsadda is very interesting, as it most closely resembles that of Kalako-deray, the most northerly and remote of the Northern Valley sites from which animal remains have been studied, and falls between Kalako-deray and Loebanr III. Table 7.4 gives an indication of the relative proportions of these main animal types at each site, both with regard to NISP counts and the MNI estimates.

The comparison of NISP counts for sheep/goat and cattle show that the Bala Hisar and Kalako-deray have the most similar figures, being the lowest difference between the two animals of all the sites. Given that Kalako-deray is the most remote (furthest from the main Swat valley, and furthest north) of the Northern Valley sites, in addition to being situated on a hill top rather than hill slope, and relatively inaccessible, the similarity with the Bala Hisar, situated on an open plain, is interesting. It suggests that there is no immediate trend in terms of rural or urban character for the animal bone assemblage. Rather, there is a trend which shows that on all the sites discussed, cattle bone comprises the greater part of the recovered and identified bone. Again, taphonomic issues may account for this trend.

The estimated MNI figures for each species and site show even more variation than the NISP counts. Sheep/goat exceed cattle at Kalako-deray and Loebanr III, yet cattle outnumber sheep/goat at Aligrama and Bir-kot-ghundai. Only at the Bala Hisar is there no difference between the two groups, and if the MNI estimates from Phase A only are considered (see Table 7.3), being the phase with recovered and identified animal bone closest to Swat PIV, then there is still an equivalent MNI estimate for cattle and sheep/goat. The sheep/goat and cattle MNI calculations could be of importance in an attempt to determine a fundamental difference between urban and rural economies on the basis of animal remains. However, they too show that in terms of the main animals

Table 7.3 Animal Bones from the Bala Hisar of Charsadda

Phase	Taxa	Number	%	MNI	Adult*	Juvenile*
A	cattle (*Bos indicus/taurus*)	3	60	1	1	-
	sheep/goat (*Ovies aries/Capra hircus*)	1	20	1	1	-
	deer (*Cervus* spp.)	1	20	1	1	-
B	cattle	24	47	3	22	2
	sheep/goat	15	29	2	12	3
	buffalo (*Bubalus bubalus*)	8	16	1	8	-
	pig (*Sus scrofa* sp.)	3	6	2	2	1
	deer	1	2	1	1	-
C	sheep/goat	1	50	1	1	-
	buffalo	1	50	1	1	-

* refers to the age of the individual elements rather than individual animals

Table 7.4 Proportion of sheep/goat to cattle at the Northern Valley Sites and the Bala Hisar of Charsadda

Site	Cattle (NISP)*	Sheep/Goat (NISP)*	Difference
Aligrama**	65	21	15 (% more cattle)
Bir-kot-ghundai	70	15	55 (% more cattle)
Kalako-deray	55	40	15 (% more cattle)
Loebanr III	60	33	27 (% more cattle)
Bala Hisar**	48	30	18 (% more cattle)

Site	Cattle (MNI)^	Sheep (MNI)^	Difference
Aligrama**	16	11	5 (more cattle)
Bir-kot-ghundai	46	32	14 (more cattle)
Kalako-deray	5	9	4 (more sheep)
Loebanr III	22	48	26 (more sheep)
Bala Hisar**	4	4	no difference

* % of total assemblage
** figure for all phases, averaged
^ absolute number estimates

represented, there is little difference between the rural and urban site animal bone assemblages. In fact, there is greater variation between the sites from Swat with respect to the relative proportions of caprine and bovine species, than there is between the Northern Valley sites, and the Bala Hisar assemblage from the Vale of Peshawar.

As one of the pig elements belongs to a juvenile, a comparison with the material from Bir-kot-ghundai and Loebanr III could be made. At Bir-kot-ghundai the age at death for pigs was calculated as earlier than at Loebanr III, giving rise to Compagnoni's suggestion regarding the different eating patterns reflective of social status at the two sites (1987, 140-1; see Chapter 6). However, given the very low proportion of juvenile bones recovered and identified in total from the Bala Hisar, it would appear that animals slaughtered while young are not significant, or not significant in this part of the site. It may be that pigs were perhaps scavengers of domestic waste, or perhaps they were crop pests, and killed as such, and having been killed, they may then have been utilised as food. Reitz and Wing suggest that the skeletons of commensal species are more likely to be intact in archaeological situations than food species used for food (1999, 204). Roberts notes that wild pigs and boar thrive today in areas of cultivated sugar cane (1987, 235), and the sugar cane grown in NWFP provides both excellent cover and a source of food for the animals. However, it was not possible to determine whether the material recovered from the Bala Hisar represented wild or domestic species.

Deer are the only known wild species for which material was recovered. The elements are from a small species, thought likely to belong to the *Muntiacus* genus (see Roberts 1987, 243), and a distal metacarpal fragment and a complete metatarsal have been identified. Each bone was recovered in different trenches, yet they are both from the right side of the body, so it is not clear whether more than one animal is represented. The specimens are from adult animals, and so may suggest that hunting was a known activity.

Age at death was calculated from the identified bones only, and these were divided into cranial and post-cranial categories for assessment. Within the cranial material, which comprised a total of thirteen pieces, three were fragments of mandible for which age assessments could not be made, and ten were teeth from cattle, sheep/goat, and pig, for which an age was given on the basis of tooth wear (see Appendix 1) (Hillson 1986; Payne 1973). The majority of teeth in this assemblage appear to have come from adults, but not elderly animals. In terms of economic decisions, this suggests that the animals in question were being slaughtered before becoming the object of diminishing returns. They had been useful for purposes other than meat, perhaps wool, dairy, traction and transport, but before losing value through age, and no longer providing the same return in terms of milk, or strength, they may have been turned into meat. Davis has suggested that to gain the maximum benefit in terms of meat, slaughter is best carried out when specimens have stopped growing at the end of the juvenile period "and the meat gain no longer increases relative to fodder input" (1987, 39).

The bone fusion data also show that age at death for both the cattle/buffalo and the sheep/goat was most likely to be adult. The categories used here to categorise the material were: epiphyses unfused (juvenile), epiphyses fused but visible (young adults), epiphysial fusion complete (adults) (Silver 1969). Of the 44 bones assessed, 41 (or 93%) were adult, three were juvenile, and there were no specimens from the *in situ* archaeological contexts that had bone falling into the 'young adult' category. Of the juveniles, two (or two thirds of the bones) were from cattle. Although it is difficult again to draw any conclusions from such a small data set, the presence of more young cattle than other juvenile animals is interesting, especially when compared with the age at death data for buffalo. All the bones identified as buffalo were from adult specimens, and while this is consistent with a purpose related to secondary products such as traction and milk, this would also, on the basis of the ethnographic data, have appeared a reasonable prediction for cattle here as well. However, young animals are not common in the recovered assemblage. When compared with the data from the Northern Valleys, this ratio of adults to juveniles most closely resembles that of Loebanr III, and if the interpretation of the age data by Compagnoni (1987, 140-1; see Chapter 6) is accepted, then this would mean that the inhabitants of the Bala Hisar had unrefined, or less refined taste in foods than those of Bir-kot-ghundai.

This interpretation by Compagnoni is queried for a number of reasons, not least of which is the small sample size used to make his calculations. Furthermore, Patel (1997, 107-8) found that calculating survivorship curves for the main domesticates at Dholavira, an urban Harappan site in India, showed that within a single city, different animals were slaughtered at different ages. Patel suggested that this may be the result of differential access to resources, and thus a function of the social organisation of the city, but it certainly indicates that there is likely to be a range of data within complex urban sites. It may be that at the Bala Hisar, young animals, just like prime cuts of meat (discussed below), were being allocated to specific recipients, and so ending up in very specific parts of the city. Without extensive open excavation across a much larger area of the mound and suitable data recovered from such an area, this suggestion is difficult to test.

Another possible explanation is that young animals are not being deliberately slaughtered and those appearing in the archaeological record may be animals dying from natural causes. In Chapter 8, which includes ethnographic information from both the sedentary agriculturalists and the transhumant groups wintering around the Bala Hisar, it will be seen that while the former keep no sheep/goat on their farms, the latter do bring large flocks of sheep/goat to winter in this area, but these are kept away from the urban areas to graze on open land. The mobile pastoral groups also reported that while lambing might occur either in the plains or up in the valleys, depending on timing of their movement between the two areas, there was very little in the way of deliberate slaughter of lambs or kids (or indeed of animals of any age, until they were past their prime) as they were mainly kept for dairy and wool purposes. This does not explain what happens to male lambs and kids (roughly half the birth population), but these may be kept to adult age for wool, or if killed, are likely to be killed while very young (Davis 1987, 39). If killed while young, they may not be viable in the market for lamb or mutton, and so consumed by the herders. This is certainly one of the specific areas of animal husbandry that needs further ethnographic and archaeological investigation in this region.

Butchery marks are reported by Compagnoni as present on 'most' of the bones examined from the Swat sites, and this is taken by him as clear evidence for the animals being foodstuffs (Compagnoni 1987, 131). However, Compagnoni does not quantify this, and does not explore the type of butchery marks further, and cut marks, or other marks on bone can be the result of a range of activities, not all related with food procurement (Hesse & Wapnish 1985, 86-7). From the identified bone assemblage at Charsadda, 3% of bones had knife marks (cutmarks and slicing) and 4% showed signs of burning (see Appendix 1). This is a relatively small proportion, and of the bones classified by size, 5% showed butchery marks, and 6% signs of burning, which is only slightly greater.

Sorting both the recorded identified skeletal parts and the unidentified parts into approximate size and element shows some trends with regard to the recovery of skeletal elements. As elements for some of the main species recovered and identified from the sites of Kalako-deray, Loebanr III, Bir-kot-ghundai and Ghaligai are also available (see Appendix 3), some comparisons between these Northern Valley sites and the Bala Hisar are also possible. Following the element groupings used in Table 7.5 (after Thomas 1986b, 134), it is clear that the majority of cattle bones recovered from the three trenches at the Bala Hisar are either teeth, or are lower limb bones (from carpals and tarsals to phalanges). The majority of buffalo bones are from the distal limbs, whereas sheep/goat are represented by equal numbers of proximal and distal limbs than other elements. In the unidentified but classified group, the majority of fragments are from the shafts of long bones, followed by fragments from the articulations of long bones.

Taphonomy, sampling strategy and collection practice are all likely to have contributed to bias within this analysis. Research on survival of different skeletal elements has clearly demonstrated that teeth, being composed of mainly dentine and enamel are remarkably tough, and that the lower limb bones such as the astragalus and calcaneus, being compact, dense bones are less likely to splinter than larger, more slender bones such as ribs (Reitz & Wing 1999, 112, 184). Certain bones such as the tibia and femur that are associated with specific cuts of meat are likely to be exposed to greater butchery than the bones of the lower limbs (Hesse & Wapnish 1985, 44-5).

With due consideration to the taphonomic and sampling issues that are likely to have biased this assemblage in favour of the lower bones and teeth, it is possible to suggest that there may be other reasons for this evident trend in the

Table 7.5 Representation of Skeletal Elements (Grouped) from the Identified Animal Bones from the Bala Hisar of Charsadda

Trenches VI & VIII						
	Cattle		Buffalo		Sheep/goat	
Skeletal Group	no	%	no	%	no	%
head (including teeth)	10	37	-		2	13
axial	4	15	-		2	13
scapula/pelvis/rib	3	11	1	11	-	
proximal limb elements	4	15	1	11	6	37
distal limb elements	6	22	7	78	6	37

recovered skeletal elements. The selection of specific body parts for specific purposes is well known in the ethnographic literature. For example, in his study of a Middle Eastern Druze group, Grantham noted that the "Druze make specific meat dishes that have a relative value of desirability from specific parts of the animal" (1995, 75). Further, sharing food with visitors plays a significant role in social and business activities, and with regard to meat, both "the choice species and carcass parts served also has symbolic meaning and communicates unspoken messages" (ibid., 75).

During ethnographic interviews around the Bala Hisar (Chapter 8), some informants confirmed that certain parts of animals are considered suitable for gifts. The head of a cow is thought to be appropriate to give to the head man and family of a village. Also, many animals are shared communally between extended families, which may comprise of up to three or four adult males and their wives and children, plus other adult blood or in-law relations. During the Muslim festival of Eid Mubarak the slaughtering and sharing of animals is expected, with the wealthier families opting for larger animals such as buffalo, and the less wealthy for sheep and goats. All possible parts of the animals are utilised, including the organs, or offal, the head and the lower limbs. Here also food, particularly meat, can play an important cultural and symbolic role (Insoll 1999, 107).

Within families, very specific parts of animals go to certain family members, in order of seniority, or other status indicators. Those farmers interviewed in the Vale of Peshawar were all Muslim, therefore the distribution and selection of meat will be governed by the Islamic dictates relating to diet and social order (Insoll 1999, 94-5). To relate such rules directly back to a pre-Islamic period is therefore impossible, but it is likely that the Late Bronze and Early Iron Age occupants of the Northern Valleys and the Vale of Peshawar had their own dietary customs that shaped the slaughter, butchery and distribution patterns. As Reitz and Wing note when dealing with archaeological assemblages the "problem is determining which elements are 'valuable' or 'prestigious', to whom, and under what circumstances" (1999, 204). While it is possible to begin to suggest that certain elements dominate this assemblage, and that this may indicate decisions that are based on social and ideological rules in place at the Bala Hisar, without a larger assemblage that has been obtained through more rigorous sampling, this remains tentative.

Within the main species identified, there is a range in the skeletal parts recovered. Buffalo are almost entirely represented by lower limbs (78%), and this trend may be partly accounted for by the robusticity of these bones. For cattle, the majority of identified bones was assigned to the head including teeth category (37%), followed by the lower limb group (22%). However, the other skeletal groups are also represented, which suggests less disarticulation of cattle carcasses. Interestingly, the sheep/goat element counts show some polarity in the representation of recovered and identified body parts. Proximal and distal limb bones account for 37% each of the remains, and head and teeth, and axial elements account for 13% each. Yet there were no identified elements at all from the scapula, pelvis or rib group.

While analyses such as the representation of body parts and MNI estimates clearly do not reflect original herd structure (Thomas 1986, 134), it may be that these types of analyses and figures can be used to make some comment about relative exploitation of animals at this particular site. The presence of a wider range of sheep/goat skeletal elements in greater relative proportions may suggest that these animals were being used for meat. In contrast, the greater number of distal limb elements from buffalo and cattle may be related to factors such as differential distribution of meat, or perhaps the different function of these animal types.

As noted above, the animal bone assemblage from the Bala Hisar at Charsadda was divided into three groups: material which was identified to element and species; that which could be classified according to size and approximate element; material which could not be either identified or classified. The second category, material which is recognisable to either element or size of animal type, but may not be assigned to a particular species, will be looked at briefly here to determine whether any major trends are evident in terms of size of animal and body parts.

In the material from both Trenches VIII and VI, larger bones, designated 'bovine' or of a size similar to the cattle and buffalo material that could be identified clearly dominates in terms of number of pieces. Material assigned to the 'caprine' category, or that of a size similar to the sheep/goat material identified, but which could also include animals such as gazelle (identified at Kalako-deray) or other small cervids, was clearly secondary. In terms of the body parts represented by each size category, from Trench VIII, most of the bovine size material comprises shaft fragments (from long bones), then articulation fragments. The smaller caprine sized category follows a similar pattern. The material recovered and classified to size and element from Trench VI is even more clearly dominated by the shaft fragment category for both animal types. This reinforces the trends evident in the material which has been identified, showing a bias in both animal types towards the recovery of long bone material.

7.2.1 Urban Harappan animal bone material

Studies of the pastoral element of subsistence related to urban Harappan sites has resulted in a recognition of many differences between them. As Meadow notes, such variability is likely to be the result of "resource availability as dictated by geographical and cultural factors" (1991, 89). While the same basic suite of domestic animals occurs on the majority of sites, and the package of winter crops is gradually replaced by summer ones (Weber 1999, 818), there appears to be little consensus about defining rural or urban sites in terms of their specific animal and plant assemblages. However, it appears that while not exclusively urban, buffalo can be considered as a possible marker for resource availability which, Patel (1997, 111) suggests, could be associated with urban, or at least settled groups. Buffalo require year round access to slow, if not standing water and mud (see Plate 8, page 374), and better quality fodder than cattle.

The presence of buffalo at Charsadda could therefore be interpreted as an indication of a settled group, with the means to supply the needs of these animals nearby. While buffalo were noted in the PV assemblage from Aligrama, they seem to have contributed very little to the overall animal bone material from either Swat or Dir. It might also be suggested that in addition to the physical restraints on buffalo keeping in the Northern Valleys, they were also largely unnecessary in an area where cattle could be used for traction and milk, and sheep/goat also for milk and meat. Buffalo were described by many interviewees as a relatively recent introduction to the Northern Valleys, and this will be discussed further in Chapter 8.

7.3 Summary and conclusions

The plant remains from the Bala Hisar of Charsadda, although limited in number are consistent in the taxa identified over a period of between 500-1000 years. Wheat and lentils are, with the exception of rice husk from the later Phase C, the only domesticated plant remains to have been recovered and identified. Both are winter crops, although with sufficient irrigation, two crops of wheat are possible per year (Thomas 1999, 306; also discussed in Chapter 8). The possible implications of a suite of plant foods dominated by winter crops include irrigation on a scale large enough to provide two crops per year to support an urban population; production of surplus and storage from a single crop; importing a summer crop such as rice from a neighbouring region such as Swat; or of course a combination of these options. Weber (1999, 823) noted wheat was the dominant cereal at the city of Harappa, during the Mature period, when the population was greatest, and the narrow range of domesticated crops is also similar to results from work at other Harappan urban sites (ibid.).

Thus far, the plant remains from the Bala Hisar of Charsadda could be interpreted as urban in nature, if the comparisons are with urban Harappan sites. In particular, the presence of wheat and the absence of barley could be interpreted as a food for a population separated to some degree from animal husbandry and agriculture. While it is likely, given the fertility and intensity of agriculture in this region today, that farming would have been occurring in the Charsadda District, it is possible that the majority of crop processing and exploitation of animals would have taken place outside the city or town boundaries. However, and this will be discussed further in Chapter 9, the role of irrigation in the fertility of the Vale of Peshawar in antiquity is not known.

A comparison of the plant remains from the Bala Hisar with those from the Northern Valley sites suggests a more complex picture of plant based subsistence with regard to urban and rural characterisation. Barley was recovered and identified from at least four of the Northern Valley sites, all of which also had wheat. Bir-kot-ghundai, which has been interpreted as developing into a town of some significance (Callieri 1992, 343), has a plant assemblage comprising wheat, rice and barley. This is a mix of both summer and winter crops, which may be the result of the conditions in Swat suited to rice production. The importance of the barley, which is the dominant plant material from this site, may be linked to the importance of cattle here, as barley straw has been noted as good for animal fodder (Weber 1999, 823). Indeed, the presence of both wheat and barley at Ghaligai, Loebanr III and Bir-kot-ghundai shows that this region was suitable for their growth, in particular wheat, which is less robust than barley in terms of growing conditions (Thomas 1999). As Weber argues that where barley and wheat are produced together, barley by virtue of its physical characteristics is the more suited to animal rather than human consumption, the barley here may well have been deliberately grown for cattle (1999, 823).

The Loebanr III plant assemblage is also dominated by barley, and again the same arguments for suggesting that it was an important animal fodder crop could apply. Rice, wheat, legumes and grapes are the other plant foods that Costantini (1987, 165) has identified here. The comparison of these two sites with the Bala Hisar of Charsadda shows some similarities, such as wheat at all three; lentils at both the Bala Hisar and Loebanr III and a narrow range of domesticated plants recovered from the Bala Hisar and Bir-kot-ghundai. Yet there are also differences between the Bala Hisar and Bir-kot-ghundai, both urban or incipient urban sites, just as there are between Bir-kot-ghundai and Loebanr III and between Loebanr III and the Bala Hisar. This suggests a conclusion that there is no major difference between urban and rural in terms of characterising the plant remains, beyond the limited number of types found on the urban sites. It could be that the location of each site may be a factor of greater influence in the selection of the crops grown in each region, and in the case of Loebanr III, this could be extended to include access to fruit such as grape, or at Ghaligai, the hackberry (1987, 156, 160).

Comparisons between the animal bone assemblages are even harder to make than with the plant material. The NISP count of the animal bone from the Bala Hisar of Charsadda is dominated by cattle in both Phases A and B, while the MNI estimate for these phases reduces the difference between sheep/goat and cattle numbers. However, these calculations do suggest that cattle may have been of slightly greater importance than sheep and goat, and certainly these two types were clearly far more important than any of the other identified taxa. Buffalo were recovered from Phases B and C. They are associated with both urban and sedentary activity (Patel 1997, 111), and it could be argued that their appearance after the earliest phase of occupation at the Bala Hisar is in keeping with the developing urban nature of the site. Deer and pig were both present in Phases A and B, and are likely to be representative of some degree of hunting or trapping around the site.

Age at death estimates show that the majority of the recovered specimens from the Bala Hisar were adult, with very little evidence for young animals becoming part of this assemblage. While Compagnoni chooses to interpret the presence of young animals as a sign of refined eating habits (1987, 141), Patel (1997, 107-8) comments that at the urban site of

Dholavira, survivorship curves simply showed that within one city, different animals were slaughtered at different ages in different areas.

The comparison of MNI estimates and NISP counts of the Northern Valley sites with both each other and the Bala Hisar showed that at both Aligrama and Bir-kot-ghundai cattle were present in greater numbers than sheep/goat, while at Kalako-deray and Loebanr III, cattle dominated the NISP counts, but sheep/goat provided higher MNI estimates (see Table 7.4). At the Bala Hisar of Charsadda, averaging the figures for all three periods gave equal MNI estimates of both animals, while the NISP counts here were most similar to those from Kalako-deray and Loebanr III. This not only reinforces the conclusions of Chapter 6, essentially that there is as much variation as similarity between the Northern Valley sites in terms of their environmental assemblages, but also shows that the Bala Hisar is similar in some respects to some of these sites, and completely different in others. For example, the Bala Hisar has a similar percentage difference between sheep/goat and cattle bone counts to Kalako-deray, but no other site has the same MNI estimate for them.

The representation of skeletal elements from the Bala Hisar does show a gross trend towards the domination of distal limbs, whereas similar analyses from the Northern Valley sites tends towards a greater overall representation of elements. This may be connected with rubbish disposal patterns in an urban area, or be the result of deliberate selection or differential access that may also be tied into an urban economic pattern.

The absence of barley and the presence of buffalo could be interpreted as the most significant factors in separating an urban assemblage from a rural package, and these two factors certainly seem to be the most significant in separating the Northern Valley sites from the Bala Hisar. However, the assemblage from Bir-kot-ghundai does call such distinctions into question and suggests that environmental conditions may also play a role in determining subsistence patterns. Interestingly, while both wild animals and plant species have been recovered from several of the Northern Valleys sites alongside domesticates, no wild material was recovered from Bir-kot-ghundai. Finally, being able to compare the environmental material from a rural site in the Vale of Peshawar from a similar time period to both the Bala Hisar and the Northern Valleys would be of great interest, and might go some way towards understanding the subsistence base in this region. As noted in Chapter 2 however, such sites have yet to be identified.

Chapter 8
The ethnographic data

The following is a summary of the main trends from the ethnographic interviews, conducted during fieldwork in both of the study regions in 1998 and 1999. Figure 3 shows the general locations where ethnographic interviews were conducted, and Appendix 5 contains a more detailed account of each interview. The main aim of this ethnographic work, and its inclusion in this study is to provide a framework within which interpretations of the archaeological material can be assessed. Modern data relating to subsistence patterns have frequently been used to develop models that can be used to test archaeological patterns in material culture (Hodder 1999, 46; Johnson 1999, 50), and the model developed here will, in the next chapter, be used to test patterns in the archaeological material analysed in Chapters 4 to 7. While the purpose of this research is to gain a greater understanding of the subsistence strategies employed by the occupants of two physically adjacent but contrasting regions during the Late Bronze and Early Iron Age periods, primarily through the analysis and interpretation of direct archaeological environmental material, the small samples from all the sites necessarily limits the scope of the analyses, and of course in turn limits the interpretations. Therefore, by incorporating information gathered through ethnographic interview and observation, a further dimension is added to the study. Identifying patterns in modern subsistence strategies in the two study regions allows a model of land use and economic organisation to be developed, and then this model can be used in conjunction with the archaeological material.

First, new material relating to the main approaches to subsistence in the two study regions will be discussed. The results of the interviews with farmers and herders who consider themselves residents in the valleys of Dir and Swat are presented first, followed by information from those farmers permanently resident at Charsadda. The material covered in the first section includes information relating to mobile groups, whether the interviews took place in Dir, Swat, around Charsadda or on the roads between the two areas. This is because they consider their presence in the Vale to be temporary, and that they do belong to, and live in, the Northern Valleys.

Second, in order to provide a critical framework in which to discuss the role of the mobile groups interviewed in relation to the archaeological evidence, a brief summary of relevant recent studies of archaeological nomadic and transhumant groups will follow the summary of these interviews. Isolating both interpreted causes of such mobility, and means by which these groups have been detected in the archaeological record, means that it possible to begin to look for similar indicators or opportunities within the material discussed here.

Third, patterns recognised in the ethnographic material will be used to develop a new model to explain and predict subsistence strategies and economic organisation in the Northern Valleys and the Vale of Peshawar of NWFP. This model will then be used to test the archaeological data in Chapter 9.

8.1 The Northern Valleys

The overall trend of the ethnographic fieldwork that relates to farmers and pastoralists connected with Swat and Dir, is one of great complexity. There are many strategies employed to exploit the land and animals, and these are related to many factors. The extremes of this spectrum of subsistence are year-round sedentary agriculture with a pastoral element permanently kept at the agricultural base, through to winter transhumance involving whole family groups, travelling from mid to upper Dir and Swat down to the Vale of Peshawar. In between there is a range of mobility patterns, and combined agricultural and pastoral approaches to subsistence. Five different major approaches have been identified through interview, and these are: one, winter transhumance; two, inter-valley winter transhumance; three, inter-valley summer transhumance; four, year-round pastoral nomadic movement; and five, permanent sedentary agriculture and animal husbandry. Figure 4 shows a stylised representation of these subsistence approaches, and indicates contact with sedentary groups in the Southern Plains.

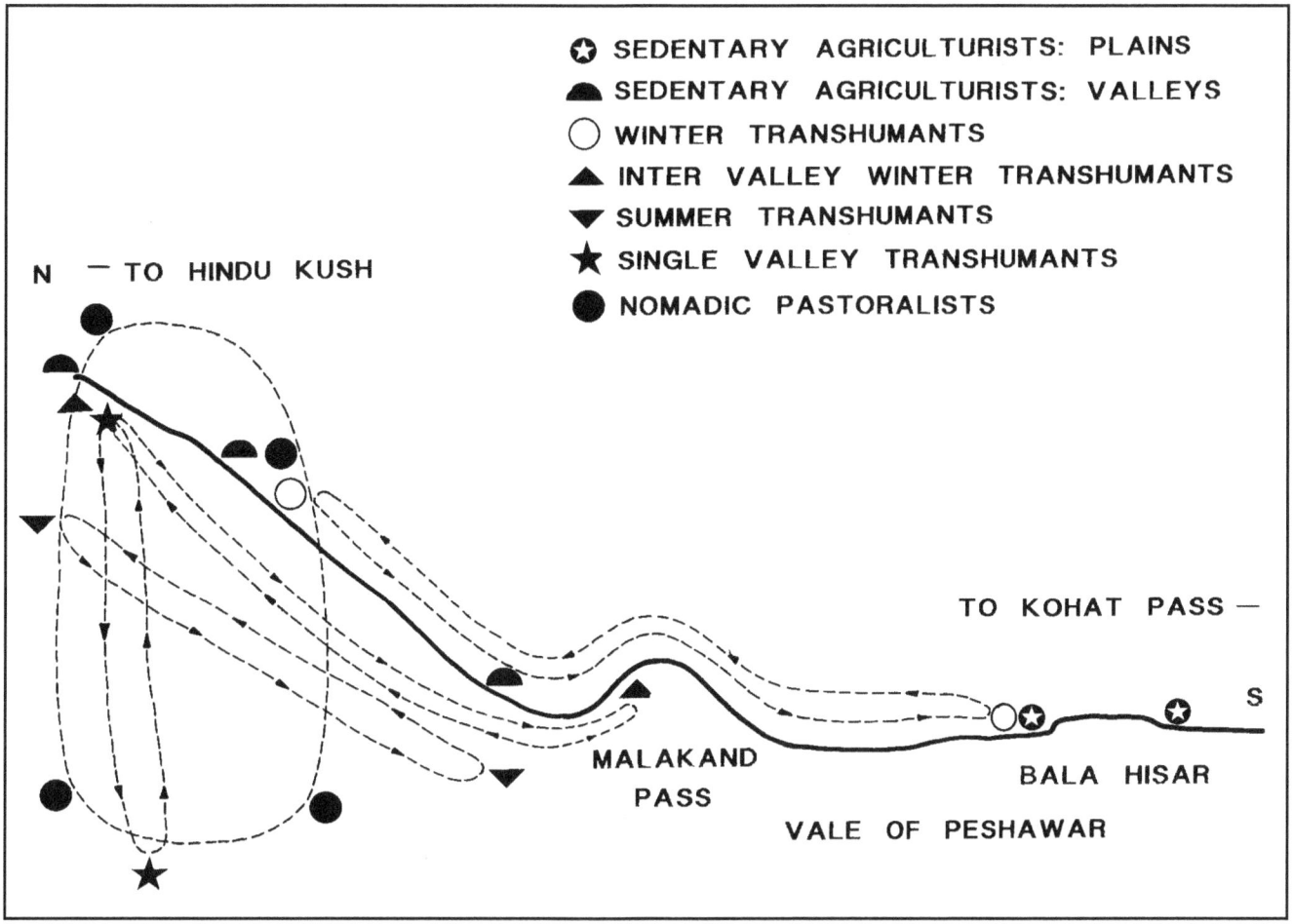

Figure 4. Cross section: Stylised seasonal movement (Alan Braby 2000)

Geography and climate do influence subsistence choices, but also important are social and ethnic affiliations, and these are undergoing continual, albeit often small and subtle, changes. The ethnographic research is an area where more than two seasons of fieldwork would be necessary to give a full understanding of the situation, however, a great deal of information has been gathered. This fieldwork was intended to establish two things: the main aspects of subsistence, such as the animals husbanded and crops grown, and the main movements (if any) of those involved in subsistence. Given the lack of detail available from the archaeological record at this stage, it seemed important to avoid superfluous ethnographic information. Instead, the questions to be asked in the field were designed with the express aim of building up a model of modern subsistence and movement, and understanding some of the main influences on these choices and selections. This model could then be used to help interpret the archaeological material, and suggest specific areas that could be the focus of future fieldwork in this region.

8.1.1 Winter transhumants

Discussion with groups in transit took place in late April and early May, spring being the time that families and other groups move back up to the Northern Valleys (Appendix 5, Interviews 1.1-1.11). These groups were almost entirely family groups, although most reported having left at least one male family member back at the 'home' up in the valleys to protect both the buildings and the crops. Other family members who were considered too sick or weak to travel any distance would also be left up in the valleys. These people on the move at this time could be divided into two broad groups.

The first of these transhumant groups, which have been designated the 'winter transhumants' on the basis of their seasonal mobility, travelled the greatest distance, and all those who came into this category talked about the history of their activities, and the established links with villages and farms down on the plains. These groups travelled from villages situated in the mid - to higher regions of the valleys of Swat and Dir down to the Vale of Peshawar. Though it should be noted here that while many of the place names given by interviewees could not be located later on maps, presumably because they were too small, or were known by a local or vernacular name, the region of their village or home was established in general through discussion with interpreters who all had considerable knowledge of this region.

The seasonal pattern was similar for all members of these groups - leaving the valleys in October and November prior to the worst winter weather, including snow. The return to the valleys was begun from mid-April onwards, but could be controlled by those who undertook paid work, such as wheat harvesting, down on the plains. This effectively meant that six months out of every year was spent away from what was described as 'home', and most interviewees said that they travelled down to the same village, usually the same house or farm every year, and had done in some cases for over forty years.

These winter transhumant groups brought the majority of their animals with them from the hills. These herds, or flocks, generally comprised primarily sheep and goat, with 50 or 60 the smallest size quoted, and the largest up to around 250. Many of the sheep were the fat-tailed variety, and while it is not possible to distinguish between fat-tailed sheep and those with normal tails (both are *Ovis aries* L.) on skeletal grounds, fat-tailed sheep are common throughout the whole study region, and indeed South Asia today. Interestingly, Kenoyer (1998, 164) in his study of terracotta animal figurines from Indus Valley sites, says that no representations of fat-tailed sheep have been recorded from figurines, pot decoration or any other art work from sites of the Harappan period. An examination of the published material from Taxila, and the artefacts from the Bala Hisar of Charsadda has also failed to uncover any depictions of fat-tailed sheep (Coningham pers. comm., 2000; Marshall 1951).

Cattle and donkey were the other two most common animals owned by these groups. The cattle were a mix of zebu, and another smaller type. The smaller type, without a hump, were described by the herders as hill cattle, and it was explained that small cattle are much better suited to grazing in very steep areas. Buffalo were also owned, but the groups interviewed while in transit between the two regions said that buffalo were transported by road, in trucks, along with elderly people. Buffalo are a recent animal utilised by those interviewed in the valleys, apparently only really becoming commonly used in the last twenty or so years. Indeed, several interviewees who kept buffalo said that buffalo milk was popular these days, and that keeping them for milk sales was profitable. Whereas in the past cow and goat milk had been adequate, with changing taste, a market for buffalo milk had emerged.

That the animals are transported between summer and winter homes by truck is interesting, as it suggests that prior to the common use of these vehicles for this trip, buffalo may not have been so widely exploited by those following a seasonal movement to the plains. Buffalo were also found to be owned by farmers who lived in the lower to mid part of the valleys, rather than those living in the higher and steeper regions, and more common in Swat than Dir. This trend is confirmed by the figures from the NWFP Agricultural Census (Government of Pakistan 1994, 861-2), which show that in total, Swat has around four times more buffalo than Dir, and that in Chitral District (to the north of both, and considerably more mountainous) only two buffalo in total were reported for the whole district.

The movement of animals is no doubt the primary reason given for the seasonal habitation change for these groups, and yet the symbiotic relationship between the transhumants and the sedentary groups in the Vale of Peshawar itself is more complex than simply a winter migration to find grazing and shelter for herds. Dung is widely used in NWFP as a building material, fuel and fertiliser. It is important to note that the demand for dung by agriculturists is great, and this means that the transhumants have a ready, ongoing supply of a valuable commodity. In fact, we were told that dung can be traded for accommodation during the winter months. Transhumants who remain down in the Vale or on the plains from October or November through to April or May will also be available for work on farms, fitting into the agricultural cycle for harvesting, particularly of the spring, or second wheat crop, or to take part in the sugar cane harvest, and subsequent gur (the raw, unrefined cane sugar) (Mian 1955, 52) production. Waste from either harvest is used as a source of animal food, particularly important for cattle and buffalo. This reinforces the mutually satisfactory relationship between seasonal herders and sedentary farmers. The transhumants can earn money, in the form of a wage for farm labouring, and the farmers have a supply of fuel, fertiliser and labour for an intensive and important part of the agricultural cycle.

These transhumants, who maintain a permanent home up in the valleys, grow crops there. They can grow one or two crops per year, the first of which is grown during the winter and the planting is carried out before leaving the valley, so the ground is prepared during September, and sown during October. The crop is then left over the winter, as the groups leave the valleys, and in many cases the ground is covered by snow for at least part of the period of absence. With the spring, which occurs slightly later in the valleys, the crops of wheat and maize ripen, and are ready for harvesting after the return of the groups, generally in May. Wheat, maize and potato were the most commonly listed crops for all the groups, although peas and fruit trees were also mentioned. Fruit crops were sold both fresh and dried, with dried fruits used by the family and transported down to the plains for sale during the winter. A crop of wheat could be grown during the summer, and some interviewees claimed they were able to prepare the land, sow and then harvest between May and October. Mainly, however, vegetables were grown around the house on a small scale during the summer months.

These winter transhumants are the group that travels the furthest in terms of distance, and whole family groups travel, with all their animals, mainly sheep and goat, and smaller numbers of cattle and donkeys. As a recent addition, some groups have buffalo that are transported by truck. The other animals continue to move by foot, following roads, and these groups often told us that roads coincided with their traditional routes. In terms of ethnic affiliation, those winter transhumants interviewed said that they were Gujars, which is interesting in that Gujars are traditionally associated with pastoralism, although whether predominantly sheep and goat, or cattle based, depends largely on the source. For example, Barth states that Gujars are both nomadic and transhumant, and herd sheep, goat, cattle and sometimes buffalo (1956, 1083), whereas McMahon and Ramsay say Gujars are nomadic, and herd cattle (1901, 16). Figures 5 and 6 show the ground plan of the home in Swat of the winter transhumants Taj Bar and family (see Appendix 5, Interview 1.8). This shows both the permanent nature of the dwelling and the provision for animals. Greater fieldwork would allow more such recording and if patterns in the recorded data were

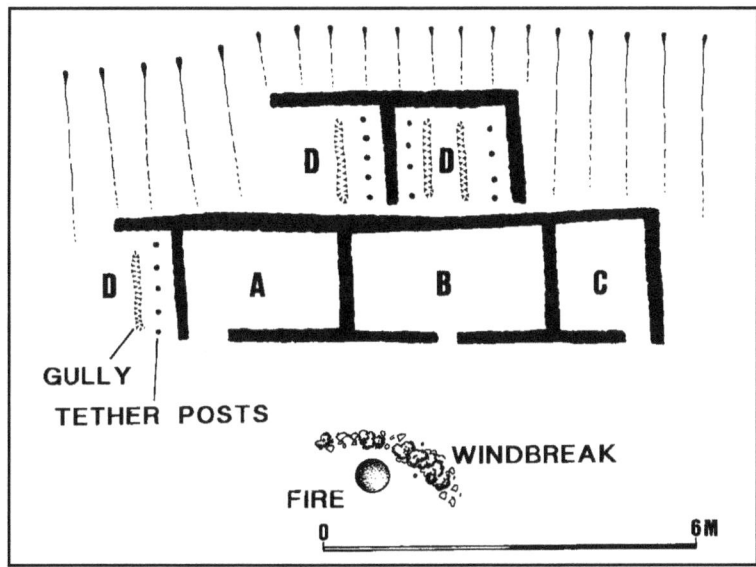

Figure 5. Ground plan of the permanent house of Taj Bar, Malam Jabba, Swat (A, B, C are living areas; D are for animals) (Alan Braby 2000)

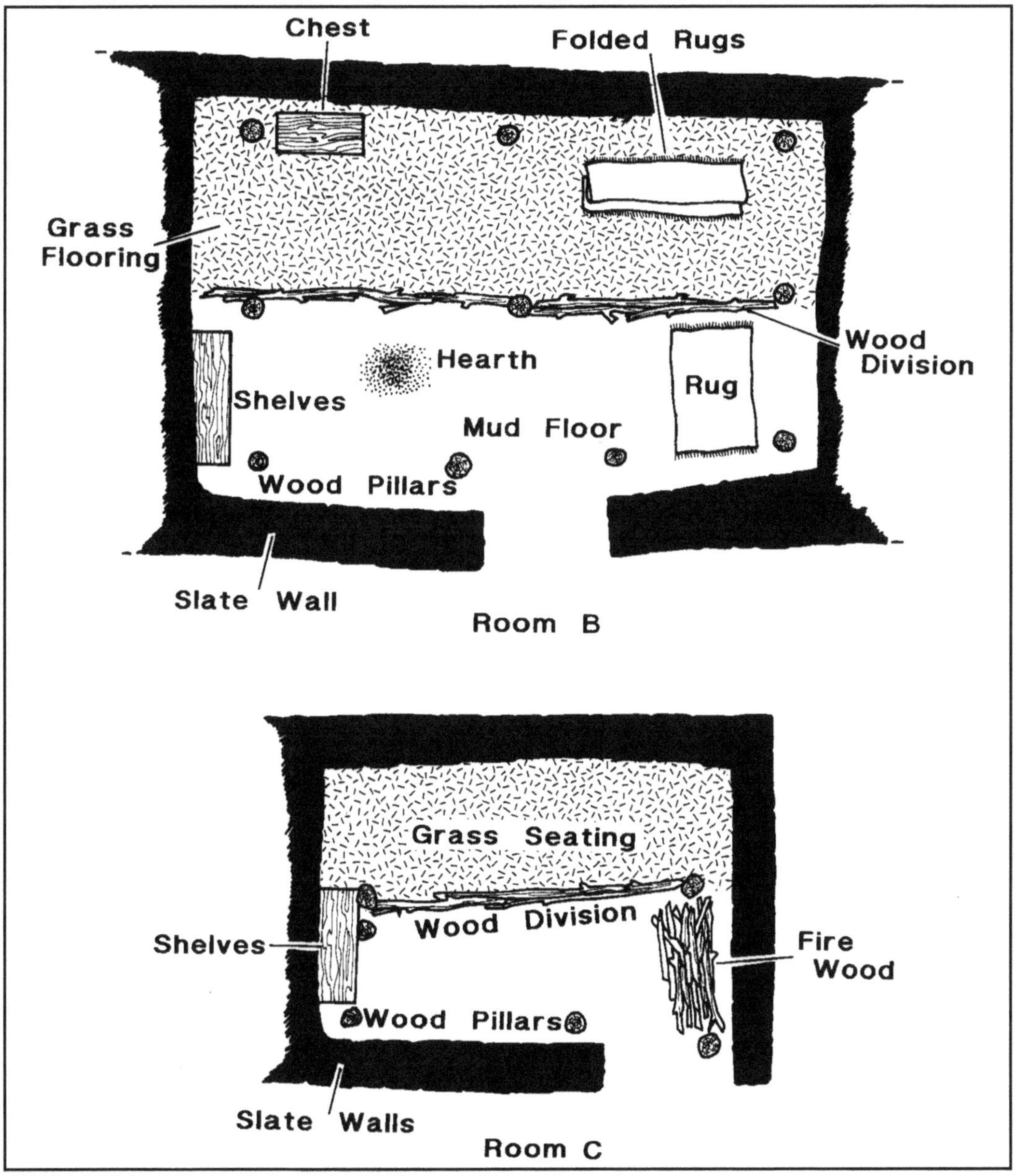

Figure 6. Ground plan of the permanent house of Taj Bar, Malam Jabba, Swat (enlargement of individual rooms) (Alan Braby 2000)

shown, then such ethnographic material could be used in conjunction with archaeological data.

8.1.2 Inter-valley transhumants

A second transhumant group interviewed also practices winter transhumance, but within their valleys of residence (Appendix 5, Interviews 2.1 - 2.4). This group of 'inter-valley transhumants' move between the upper and lower regions of Swat and Dir, and like the longer distance winter transhumants, whole family groups make the seasonal journeys. While they regard their home as the upper part of either valley, bad weather during winter means that to find good pasture for their herds, they bring these animals down to the lower part of each valley, close to the Malakand Pass area. The journeys both down from the home, and back up to it again occurred at the same times of the year as for the winter transhumants, moving down in October and November, and back up again in April. The main animals kept by families in this group were sheep and goat, with some cattle and donkey, but none were recorded as keeping buffalo. Both zebu and small hill cattle were kept. Crops were also grown at 'home' over the winter months, being mainly wheat, maize and potatoes, and in summer, vegetables.

However, there does not seem to be the same reciprocal labour and goods exchange between the inter-valley transhumants and their winter hosts, as between the winter transhumants and hosts. When questioned, fewer said that they returned to the same village or house or farm, more said that they went to different places each year. This may suggest that there is greater pressure on resources in this region, or that the things that the transhumants can offer their hosts are already available here. Conversely, the main requirement of the mobile group is grazing, and this may already be under pressure from those who live there year round.

The ethnic composition of the inter-valley transhumants is slightly more complex than that of the winter transhumants, as some of those interviewed said they were Gujars, and some said they were Pathans. Barth (1956, 1080), from his analysis of the ethnic groups of Swat, says that Pathan social organisation is based on settled agriculture, which produces a surplus necessary to support full time specialists. Further, according to Barth (ibid., 1081), only Gujars and Kohistanis incorporate regular mobility into their subsistence strategies. However, as all the transhumant groups, regardless of the type, lived on, or owned land in their 'permanent' village, this is not necessarily contradictory. A strong impression gained from the interviews, but one rather hard to quantify, particularly through interpreters, was that the inter-valley and summer transhumants (below) did not consider themselves as mobile, but rather making a purely practical decision in terms of herd management.

8.1.3 Summer transhumants

A third type of transhumance practised in Swat and Dir is 'summer transhumance', whereby families who live in the lower parts of the valleys move with their herds up to the upper parts of the valleys between the months of June and August/September (Appendix 5, Interviews 3.1 -3.4). The reasons given for this movement were both to provide good grazing for their animals, and also for the people to avoid the greater heat of the lower valleys during these months. The animals kept were sheep and goat, cattle (both zebu and small non-humped hill type), and donkeys. Although movement in this latter group is during the summer, rather than the winter period, again wheat, maize and potatoes are grown over the winter, and harvested in the spring before moving further up the valleys.

Again, whole family groups moved with the animals, but these summer transhumants contrast with the winter transhumants and inter-valley transhumants in considering the place where winters were passed 'home', rather than where summers were spent. Further, unlike the first two groups who appeared to divide their year roughly equally between the two areas, the major part of the year was spent by the summer transhumants at the house and village considered home. The summer transhumants all described themselves as Pathans and Gujars, and again this may reflect economic factors such as land owning versus renting, rather than an ethnic division between subsistence practices.

8.1.4 Nomadic pastoralists

The group of people known as Ajurs are the group who are closest to nomadic pastoral status rather than transhumants (Appendix 5, Interviews 4.1 - 4.3). Ajurs have flocks of sheep and goat, and do not have what they consider to be a fixed residence. They do have villages that they indicate an affiliation with, but they do not have a permanent house, or house occupied consistently or regularly for part of the year. Like the transhumant groups, and indeed many nomadic groups, Ajurs follow a seasonal pattern in their movements, with grazing for their flocks their prime consideration. Therefore, summer movement is designed to take advantage of grazing up in the mountains, and is also cooler and more pleasant for both humans and animals. In winter, moving to lower land avoids most of the snow and floods, although Ajurs in Swat and Dir tend to remain within the valleys rather than travelling down onto the plains. The absence of a permanent residence and the continual movement with flocks, means that Ajurs as a group do not cultivate crops. Rather, they exchange animal products such as milk, ghee and butter for plant foods (and a range of other goods).

Changes in the modern social and economic environment in northern NWFP have resulted in a change in the role and status of many Ajurs. While previously the majority of households in Swat and Dir practised subsistence agriculture and pastoralism, this is no longer the case. Now, many family groups have members working outside the household or farm, and this means that there may not be sufficient labour to undertake all tasks such as herding and care of animals. Therefore, Ajurs may now be hired to undertake the care of animals for a number of households, most frequently over the summer months. These animals are taken up to summer pastures, and ghee, milk and butter regularly sent down to the owners. The attendant Ajurs are paid according to the number of animals they care for. The summer pastures can be many hours, up to a days walk, from the village, and children are often used as carriers of produce from the pastures to the villages.

Ajurs themselves are also conscious of different social perceptions, and many are now claiming that they are not Ajurs, but are in fact Gujars, or settled people with some land and cattle in addition to sheep and goat. Ajurs believe they are looked down on and are considered inferior within modern social order. While some of those interviewed did say they had been Ajur in the past, or rather their parents were Ajur, they were adamant that they themselves were Gujar (Appendix 5, Interview 5.4, Abdul Raziq).

8.1.5 Sedentary farmers

Finally, there are settled farmers who remain with their families in the same home all year round (Appendix 5, Interviews 5.1 - 5.8). Like the mobile groups, they keep goat and sheep, and perhaps some cattle or buffalo, although the buffalo were again described as recent. Buffalo were not kept by any of the groups interviewed much further north than Mingora, towards Bahrain. As the valley gets much steeper here, buffalo would be very impractical. The same basic suite of crops is also grown by the sedentary farmers, but rice was also added to the list. Rice, although an important crop in lower and mid Swat was mentioned only by those who remained in the valley all year round. As a summer crop it would be more than possible for the winter transhumant, or winter inter-valley transhumant groups to cultivate it, although not the summer transhumants. While there is the possibility that those interviewed failed to mention it because they did not consider it significant, it may be that for some reason, perhaps related to land ownership, status or the organisation and commitment required for maintenance of the paddy system, only permanently sedentary agriculturists are prepared or able to grow rice.

Some of these farmers told us that their animals were sent away during the summer to take advantage of the good grazing up in higher mountain pastures at that time of year. They were accompanied by either family members, such as sons or nephews (ie young males) or sent away under the care of Ajur groups. In an attempt to discover why these groups were sedentary, they were asked a range of questions aimed at understanding why they did not move with the seasons, and how long they and their families and forebears had practised agro-pastoralism. The response was in general that this was simply how they lived, and how their parents and grandparents also had lived. In terms of the ethnic identity of these settled farmers, a broad division was apparent according to geographic area. In lower and mid Swat and Dir, they were Pathan and Gujar. Further north in Swat, around Kalam, the interviewees told us they were Kohistanis, or Swat Kohistanis. While Kohistan is shown on many maps as a separate district within NWFP (north of Mansehra, west of Swat), it is also locally simply used to describe mountainous regions, hence the designation of places and people as Swat Kohistan, Dir Kohistan, Indus Kohistan.

When interviewing farmers in the higher areas of Swat, such as around Kalam, it was of particular interest to find out how they coped with the inevitable snow during winter, and we were told that there was no major problem while roads and paths into the nearest village were kept open. This was achieved through a combination of government controlled mechanised means such as diggers and road sweepers, and also by community organised labour clearing away snow. During severe winter conditions, therefore, access to resources was a controlling factor, and relied on both forward planning in the form of collecting and storing animal and human food and fuel, and being able to reach external sources. Successful year round occupation of such extreme snow bound areas is also likely to have been possible in prehistory, providing access to stored food and a fuel supply was consistently available. The obvious need for surplus and organisation is worth investigating here in terms of the implications for the social organisation of the archaeological sites and their occupants, and this will be done in the following chapter.

8.1.6 Northern Valleys: discussion

This review of different modes of subsistence in the valleys of Swat and Dir highlights a number of trends. First, just as there is a range of geographical and climate divisions within the region (see Chapter 3), there is a range of ways of exploiting the land. However, these different modes of subsistence do not neatly echo the physical divisions – rather social and economic factors appear to be just as important in determining how people exploit the land. Mobility, particularly in the form of regular, seasonal transhumance is well known, and is a response both to seasonal climate, and also as a means of getting the best possible food for the animals at any given time of year. Without question, sheep and goat are the animals favoured by transhumants of whatever pattern, and while many do keep smaller numbers of cattle, only those travelling right down into the plains over winter reported keeping buffalo. The sedentary farmers also kept buffalo, again a recent innovation that was described as the result of a change in taste for milk.

The social or ethnic role of sedentary subsistence was not clear, beyond the contact with the family who were described now as settled Gujar, but had been mobile Ajar. Trying to establish how different groups consider other groups, in terms of their ethnic, social and subsistence affiliations could be of great value in future research. That a large number of families do remain in their houses in parts of Dir and Swat sufficiently northwards to be snow bound from November through to February or March, or even later is without question. This, combined with the range of seasonal movements described, suggests that above all else there is a wide range of subsistence responses that enable many social, ethnic and economic groups to co-exist in this area. Mobility is important, and has a long history in this region. Mobility has also resulted in significant and long standing contact with the plains.

8.2 Charsadda District

The following is a summary of the main trends from the interviews with a series of farmers working the land around the site of the Bala Hisar itself (Appendix 5, Interviews 6.1 - 6.12), which today could be described as urban hinterland. These farmers all lived permanently in this area, many reported that their fathers and grandfathers had worked the same land, and even for generations before that. There was no seasonal movement involved, and these farmers had one place of residence. The majority of farmers interviewed said that they did not own the land they worked on, but were tenant farmers, or share farmers. The most common arrangement was that the farmers were able to retain 50 percent of the crop, or profit from the crop grown, with the other 50 percent being retained by the owners.

In direct contrast to the subsistence strategies apparent in the Northern Valleys, it seems that making a living from the land in Charsadda involves fewer choices or decisions. The sedentary nature of the farmers and families, the potential for the land to support large numbers of people and animals all year round, and the relatively narrow (and standard) range of crops grown and animals kept, may be the result of many social and economic forces, as well as the contrasting nature of the climate and geography or the plains and valley regions. However, the result is information in the form of these interviews which gives a far more uniform account of subsistence on the plain than in the hills.

Cash cropping was a priority in this region, something that is likely to be enhanced by the nature of land ownership. Sugar cane was the primary cash crop grown by farmers in this area, and is either sold to sugar mills, or processed locally as gur, and then sold. The difference between these products is that the price of sugar is set by the government, whereas the price of gur is the result of free market forces (Aurangzeb 1989, 2-3). While those working land at the larger end of this scale claimed they were able to support their families entirely from the farm output, many reported that at least one member of their family group had to seek other employment.

Sugar cane can be processed locally, as there is a communal processing area adjacent to the Bala Hisar, which is operated by local farmers with zebu, and often employing transhumant labourers to assist, between the months of November to February. The decision to grow sugar cane is made by the land owner rather than the farmer, and many farmers said that sugar cane had been grown on their land for as long as they and their forebears could remember. Irrigation is essential for the growth of sugar cane, and Charsadda has a combination of Government and private irrigation, both of which rely on the provision and maintenance of channels. This is done by communal labour forces, and decisions about maintenance and availability of water are made by village councils, composed of village elders. Sugar cane is planted around March, and then harvested and processed between November and February.

In addition to sugar cane, all the farmers reported growing wheat for their own needs. There can be two seasons for wheat, although at least three quarters of those farmers interviewed said they grew only one crop of wheat per year. Those growing one crop said it was sown around October and harvested in April, and those with two crops reported sowing again in May and June and harvesting in September. Harvesting, particularly of larger fields, draws on labour from transhumant groups, some of whom remain down on the plains until the harvest is complete, to take advantage of the wages offered. Again, the wheat is processed locally, with the nearest mill to the Bala Hisar around one mile from the farms. This wheat is for domestic use only, with very few farms in this area being large enough to support extra wheat crops for cash. Wheat is often grown in the same field as the sugar cane, planted between the canes themselves. This requires some intensity of labour for planting, weeding and harvesting, but between the permanent labour force on any given farm and the hired in labour for specific tasks, this seems to pose no difficulty.

The only other crop that was grown with any degree of regularity around the Bala Hisar was a crop of a clover type (*Trifolium* spp.) which was used for animal fodder. This was grown as a cash crop, but only by those who had large farms. It was a good crop to grow, being very quick, and rather easier than sugar or wheat. It is also likely that the nitrogen fixing properties may have been recognised (Mabberley 1987, 590). Clover could be harvested at least twice a year, and would grow again quickly, like grass. This meant that although not bringing in such a high price as sugar cane per harvest, clover produced comparable profit given the frequency of harvesting. This animal fodder was sold in both Charsadda town and Peshawar to those living in urban areas, keeping animals for domestic or commercial use, but with only limited or no access to grazing.

These farmers spoke of other crops such as lentils and peas, but none of them grew these types of plants on their land. This was no doubt due to the economic forces that led the owners of the land to decree that only sugar cane, or in some cases the animal fodder, should be grown for cash. No-one grew rice. It was not considered a suitable crop to grow in this region, despite the irrigation. This may be due to the higher prices available for sugar, or it may be that the land itself was not suited to paddy, unlike the land in Dir and Swat. For example, in Dir, there is more than forty times the area under paddy than in Charsadda District (Government of Pakistan 1994, 407, 412).

None of the farmers interviewed kept goats or sheep. They said they had no need for these animals, and if they wanted goat or sheep meat, they would buy the animals at the market. Live animals were not often bought. The farmers said that they would buy cuts of meat for their family to eat, if they were needed. Live animals were for certain occasions, such as Eid. Chickens and eggs were a commonly cited source of animal protein, and all farmers kept at least five chickens in their compounds. Cattle, described as bullocks, and noted as almost always zebu, were kept by all farmers. They were important for milk, used for ghee, butter, cheese and yoghurt, and for ploughing. Buffalo were also kept, and again provided dairy products and traction. Dairy products were sometimes sold, if more was produced than needed by the family group, and the urban areas of Charsadda and Prang provided a nearby market. The cattle and buffalo were generally kept next to the farmhouses, and grazed during the day by children of the family, when not being used in the fields.

8.2.1 Charsadda District: discussion

In terms of the ethnographic material, it is very clear that there is a far greater degree of uniformity in the modes of subsistence around the Bala Hisar, in what could be described as an urban hinterland area in the Vale of Peshawar, compared with the Northern Valleys. While this can in part be attributed to economic forces, the nature of the such factors as the topography and soils are also clearly important, and has largely helped shape the modern economic demands and approaches. While both lower Swat, and to a lesser extent Dir, are fertile, the fertile ground is limited, and this seems to put greater emphasis on the need to integrate animals, particularly goats and sheep, into subsistence. Today, there is a much greater range of agricultural crops grown, including fruit crops, in the areas of the sites discussed in Chapters 4 and 6. This may represent a risk spreading strategy, or one that is aimed at taking advantage of a fertile area, but one subject to climatic stress.

In the Vale of Peshawar however, which is an area of Pakistan noted for its fertility in both modern and historic times (Wheeler 1962, 1), people seem to have a far less varied approach to subsistence. The dedication of farms to growing a single cash crop and a single food crop of wheat, is indicative of the reliability of crops and climate in this region. The water necessary for buffalo is always there, and the good grazing for quality milk from buffaloes and cows is available all year round. The occupants of the Vale of Peshawar have no need to move away for half the year to find food for their animals and to shelter from harsh winters. They are sedentary, and as such can grow crops that require year round attention, and harvest and process these crops.

8.3 Transhumance in ethnoarchaeology and archaeology

The causes and origins of transhumance as a strategy for the procurement of pasture for herds through seasonal movement in prehistory has resulted in a number of studies from various parts of the world. Identification and analysis of explicitly transhumant groups do not appear in the archaeological literature to the same degree as nomadic groups, but some European studies include Lees and Bates (1974) work, where the origins of transhumance are considered, and strong emphasis is placed on unequal power within a reciprocal exchange network of mobile and sedentary groups. Chang and Tourtellette (1994) used their ethnographic studies of transhumant groups in Greece to build up a model of the artefactual and structural remains that could be expected in archaeological situations. They stress the importance of structural remains, and of looking for a whole range of data types.

In his discussion of transhumant economies in the Pindhos Mountains of Greece, Halstead says that "claims for transhumant pastoralism in prehistory are ultimately based, implicitly or explicitly, on analogy with recent mountain economies of the Mediterranean" (1990, 61). He goes on to point out that there is a lack of direct evidence of transhumant practice in archaeological studies, and further, that much of the wider Mediterranean practice is the result of recent (historically recorded) economic and environmental change that cannot be transferred back into prehistory (ibid.; 1987, 79-80). Halstead concludes by suggesting that while the archaeological evidence for the earliest prehistoric settlement in the Pindhos region (c. 1000 BC) supports an economy based on mixed farming, an element of pastoral mobility is quite possible, yet such groups may well be archaeologically invisible (ibid., 72). However, Halstead also notes that in many areas around the Mediterranean, modern practices indicate very complex economic and social factors behind the traditional subsistence strategies, and says that this very complexity is of great significance in determining the types of information that we should aim to gain from archaeological data (1987, 86-7).

Gilbert (1983, 107) also recognised the range of subsistence strategies apparent in the archaeological record from western Iran, where transhumant, pastoral nomadic and sedentary agriculturalists have been demonstrated as successfully exploiting different strategies and also maintaining social and economic links. Analysis of archaeological remains from the plains in the Khuzistan steppe region of south western Iran dated to c. 6000 BC, indicated a dominance of the faunal assemblage of sheep and goat, and within this, an age at death curve almost entirely lacking young animals (ibid., 108). General conclusions from this study were that while there were great difficulties recognising contact between sedentary and mobile groups on the basis of animal remains, where an assemblage is so completely dominated by one taxon, indications of specialised pastoralism are strong. Further, that an assemblage lacking in young animals is consistent with the division of a herd, on the basis of function, where breeding, elderly, and sick animals may be kept apart from healthy adults (ibid.).

In terms of the visibility of nomadic groups in the archaeological record, nomadic camps are more visible that transhumant temporary, mobile camps, but there are nomadic seasonal camps that may be useful for comparison. Cribb (1991, 92-6) gives a list of features of nomadic camps, from structures through to faunal material that he says are indicative of nomadic site occupation. Faunal material is considered difficult to deal with in terms of reconstructing herd patterns (Cribb 1987, 377), but there are various theories about detecting herd management strategies through animal bone assemblages in the general archaeozoological literature (eg Payne 1972; Halstead 1992). It is recognised that while organic material can provide direct seasonal indicators, such as the presence or absence of seasonally available species, and the analysis of growth patterns in bones and mollusc shells, features and artefacts can also play an important role in interpreting seasonal activity in past communities (Cross 1988, 55). For example, site location itself and storage pits can contribute to the understanding of site occupation. However, the use of a number of indicators together is considered the most effective way of approaching an analysis of seasonal activity (ibid., 58).

The results of extensive ethnoarchaeological analysis of mineral residues in manure and burnt dung, rock polish from the walls of shepherd's caves, as well as structural remains associated with herding such as stock pens have been tested using archaeological data from sites in the western Mediterranean (Brochier et al. 1992). The results from this study suggest that the use of caves and other stone structures in association with herding was practised as early as the beginning of the 5th millennium BC in this region (ibid., 90).

Within the subcontinent, Rajan's (1994) study of site distribution during the Iron Age and megalithic period of Tamil Nadu in southern India investigates the subsistence base of the site occupants at a regional level. Earlier researchers in this area such as Narasimhaiah (1980, cited in Rajan 1994, 137) suggested that such communities were mobile pastoralists, even nomadic pastoralists, and indeed that the builders of the megalithic monuments were an entirely separate group from the occupants of the habitation sites. These simple models have been discounted by Rajan, but the role of regular, seasonal transhumance does not appear to have been fully explored within this area of contrasting environmental zones (ibid., 137-8).

Mughal (1994, 53) carried out an archaeological survey in the Cholistan Desert in Punjab, Pakistan, which identified a significant number of nomadic sites dated to the Hakra and Harappan periods. These sites were divided into two types, being either temporary camp sites, or more permanent sites used intermittently (ibid., 58). From an analysis of the size and number of sites, Mughal has been able to distinguish changes over time that relate directly to formation and development and decline of the Harappan Civilisation in this region (ibid., 61).

Within NWFP itself, Thomas (1983, 1986, 1999) has made use of analogy with modern subsistence practice to help interpret archaeological environmental material from sites in the Bannu area. Thomas has identified patterns in the environmental material from sites dated to between 4240 - 2040 BC (1999, 312, 314-5) that relate to exploitation of both plants and animals, and then presented a range of possible interpretations of the material in terms of economic organisation.

The importance of transhumance in prehistory is clear from these studies, and the recognition of mobile subsistence strategies in South Asian archaeology has facilitated the interpretation of archaeological material. One of the limitations of this study is that this section is ethnographic rather than ethnoarchaeological in nature. This means that it has focused on trying to understand more about the ways in which the inhabitants of Swat, Dir and Charsadda exploit the land for subsistence and economic purposes. Through direct interview the intention was to understand more about traditional lifestyles in terms of economics and subsistence, and then to use this information to suggest possible interpretations of archaeological material.

In contrast, Middle Range Theory is a way of "assigning meaning to empirical observations about the archaeological record" (Bettinger 1991, 62), and as such has utilised ethnoarchaeological studies as a means of linking 'static' objects in the present to the 'dynamic' processes in the past that created them (Binford 1989, 167, 239; Johnson 1999, 49-50). According to Hodder, Middle Range Theory is crucial to Binford's work as it "provides the link between statics and dynamics because it describes and understands the formation of the archaeological record, using general theories about use, discard and post-depositional processes" (1999, 27). This work is not ethnoarchaeological as it did not have as an objective, and so did not record the specific details of material culture in the agricultural and pastoral groups that could then be considered as archaeological indicators. Rather, the ethnographic work that has been conducted was intended to provide general subsistence pattern lifestyles that can be tested using archaeological material, and further, indicate the type of analyses and interpretations that could be possible in future work, given a suitable environmental data set. Recording and analysing the material remains of transhumant groups to build up a model that could then be transferred to interpretations of the archaeological record, would be a logical and possible way to progress this work, but would require considerably more fieldwork, and access to a range of different types of transhumant groups throughout all the seasonal moves they made. However, the potential for carrying out ethnoarchaeological work is indicated by Figures 5 and 6, which show the ground plan of Taj Bar's permanent home in Swat, which shows Taj Bar's animal shelter, or area D in Figure 5. Recording other such houses, and other structures associated with seasonal transhumance would then allow comparison with archaeological structural material.

8.4 Patterns in the ethnographic record: developing a model

Patterns of subsistence strategies in the two regions studied can be seen in the ethnographic material. Although complex, the various mobile approaches of groups in the Northern Valleys do follow regular seasonal patterns, and these can be mapped (see Figure 4). Further, the sedentary groups of both regions can also be described in terms of their subsistence activities. When the different defined strategies are put together they result in a dynamic model of interaction and contact.

This model is depicted in Figures 4 and 7, and is based on the subsistence categories defined in the ethnographic interviews. Permanent sedentary groups are shown for both regions, in the urban area of Charsadda, and its hinterland, likewise in Swat and Dir, at the urban sites, and also in the surrounding countryside. It is important to remember that if sites such as Bir-kot-ghundai were urban and incipient urban

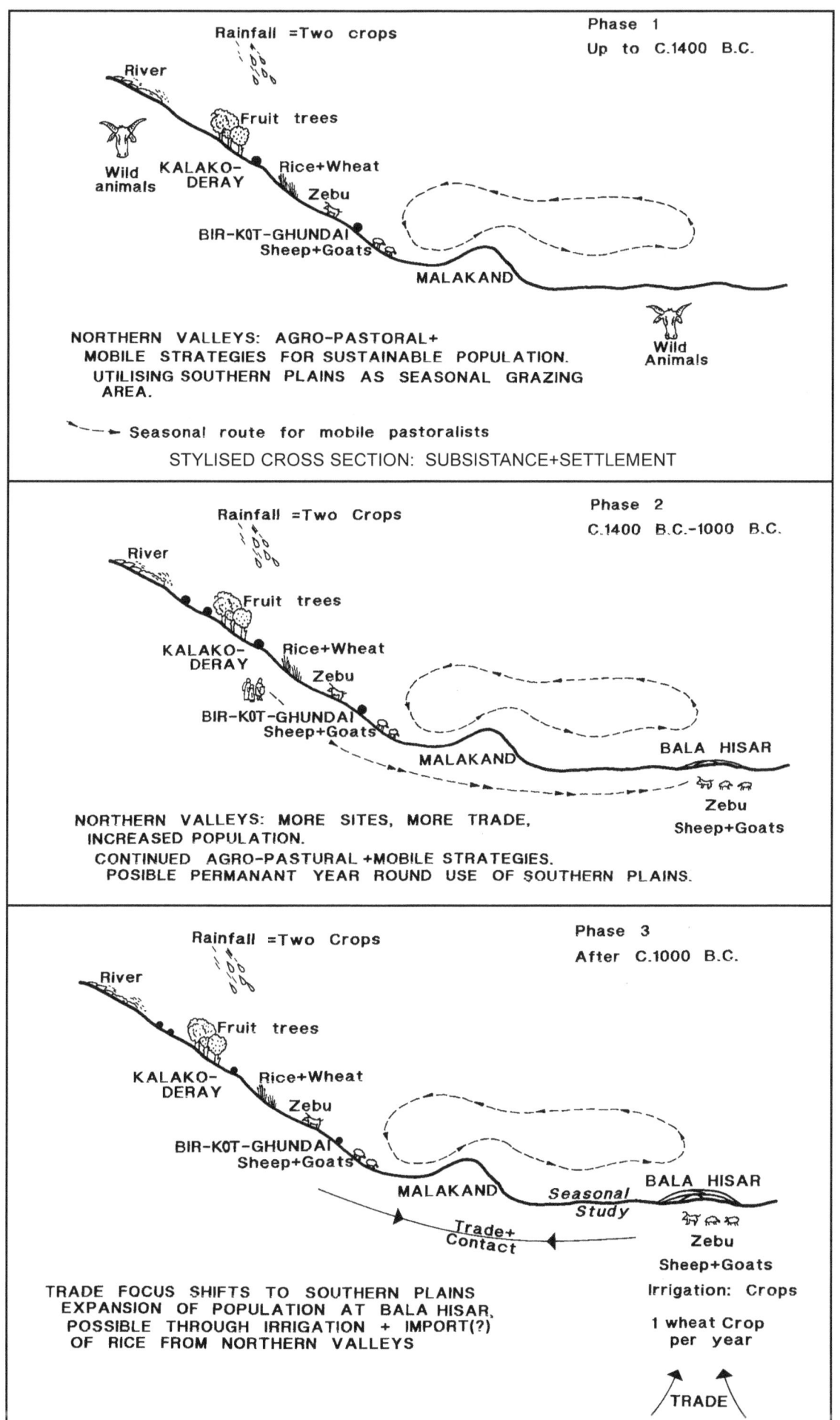

Figure 7. Three phase model of subsistence and settlement development (Alan Braby 2000)

in nature during the prehistory of Swat, then they are likely to have had a permanent population, regardless of the surrounding rural subsistence strategies. Permanent groups in the northern or upper areas of both Dir and Swat, for example at Kalam, need some form of storage for winter human and animal food. Rooms either within houses or adjacent to animal shelters are used for this, but it is possible that the so-called dwelling pits at many of the Swat sites could have served a storage function. In the Northern Valleys, the range of mobile strategies linking both upper and lower valleys are shown, and also those linking the valleys and plains regions. Movement of people, animals and other goods, including crops such as rice and fruit, are shown in Figure 7, with arrows giving the direction of this movement.

The crops and animals contributing to the main subsistence base of each group, as described in the interviews, are also shown, and from this it can be seen that there is considerable overlap in terms of the types, particularly in the Northern Valleys. This has implications for trying to distinguish the different groups on the basis of the environmental remains. This issue has been explored in a study of the subsistence base of the Kalasha group located in the Chitral Valley to the north of Dir. While Parkes (1987) claimed that ideological beliefs were behind subsistence decisions made by the Kalasha, more recent work comparing the Kalasha subsistence with that of their Muslim Kho neighbours, found very little difference between the two groups (Young et al. 2000). Rather, it was suggested that the marginal environment of the Chitral Valley is likely to shaped these decisions.

However, a comparison of the Charsadda District farmers with the Northern Valley groups does show some major differences in terms of the animals kept and crops grown. Again, there is an opportunity for overlap between these types when the mobile groups make contact with the plains groups, and it is this mobility which is one of the keys to the model. Barth, in his paper discussing social stratification in Swat, says that while "Swat lies in the middle of a turbulent cultural shatter zone, it is geographically isolated in that no major routes of communication pass through the valley" (1969, 115). While this may be the current situation, it is by no means one that is supported by historical or archaeological evidence. Here it is argued that the regular seasonal mobility from the Northern Valleys to the plains region supports a model of contact, and in addition to such movement being a valuable subsistence strategy, it also plays a role in cultural development. When seasonal subsistence related movements and trade routes are shown together, the result may be similar to the stylised sources of influence shown in Figure 8.

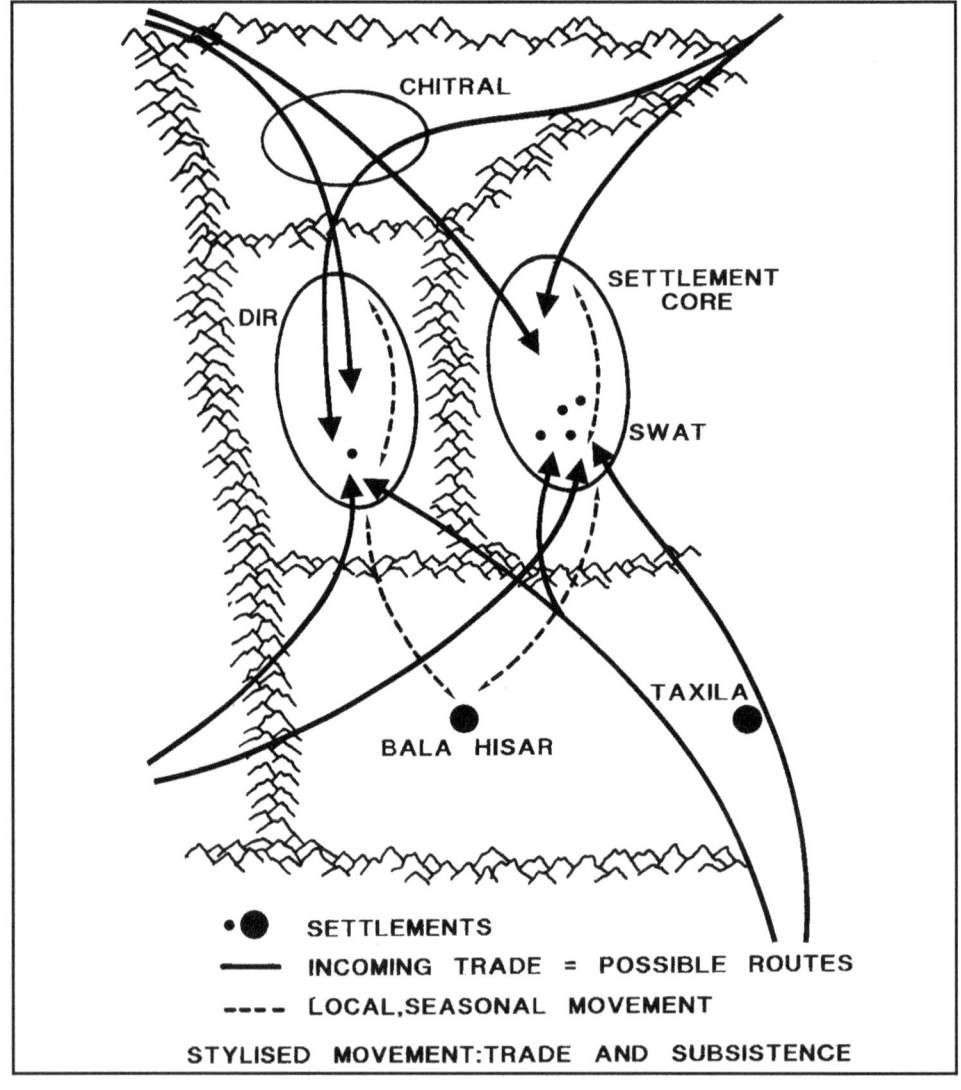

Figure 8. Sources of influence in the northern valleys & southern plains (Alan Braby 2000)

8.5 Summary and conclusions

The ethnographic interviews show one very marked overall trend: subsistence patterns in the Northern Valley areas are complex and multiple, while in the Charsadda District of the Vale of Peshawar they are comparatively very simple. This is of great interest and importance, because while the geographical elements of these two areas are in great contrast, and it could be argued that contrast in subsistence between the two study regions would be expected, there is a far greater range in terms of approach within the single region of the northern valleys. There is the expected contrast between plain and valleys, but the enormity of the contrast between different groups within the valleys themselves is more striking.

Barth's analysis of the different subsistence strategies in Swat was based on the division of the population into different ethnic groups, that occupied distinct if closely linked 'ecological niches' (1956, 1079). While this new study suggests that different ethnic groups do indeed tend to have different subsistence strategies, this is by no means exclusive, and by no means closed to change. However, it is clear that in the Northern Valleys, different strategies are tied to many factors including ethnic group, social standing, family habit, and also feeding animals. Given that all the groups discussed above are operating within the same environmental and geographical constraints, they represent a range of very different means of subsisting, from avoiding harsh winters altogether, through to remaining in one place and surviving winters by storage and other contact.

In the following chapter the available archaeological data will be tested and archaeological trends predicted. The different approaches covered here will be considered in relation to the archaeological material, both environmental and other, and the results critically discussed. Further, the means of assessing evidence for different subsistence strategies in the archaeological record will also be discussed. At this point, it should be noted that the material suggests that a range of subsistence options are known and practised in the Northern Valleys, while in the Vale of Peshawar, agricultural and pastoral activities are concerned as much with cash cropping and providing animal food for urban areas, as with producing food for the immediate needs of the family group. As has been noted in the archaeological environmental data, the range within the Northern Valleys is perhaps greater than between valley and plain.

Chapter 9
Discussion

In this chapter the conclusions about the nature of the environmental evidence from the sites in the Northern Valleys and the Bala Hisar of Charsadda, and the supporting archaeological material will be discussed, and this material will then be tested using the model developed from the ethnographic research presented in Chapter 8. Prior to testing the new model using the environmental and archaeological data summarised and presented in Chapters 4-7, five extant interpretative models or theories relating to the economic organisation of the sites discussed in this volume will be summarised, and their relevance to the environmental material itself will be assessed. These models and theories cover both overlapping and opposing theoretical views in an attempt to build up a picture of the current interpretative approaches and how they are applied in this area. Particular attention will be paid to elements of the economic and subsistence base of these models and theories, if and when they are included.

These models have been selected because they are based on excavation information from the sites of Balambat, Timargarha, Aligrama, Bir-kot-ghundai, Ghaligai, Kalakoderay, Loebanr III, and the Bala Hisar of Charsadda. Professor Ahmed Dani of the Quaid-I-Azam University, Islamabad, formerly Professor of the Department of Archaeology, University of Peshawar, (1968, 1992) directed excavations at Balambat and Timargarha. Professors Giorgio Stacul (1996, 1994a,b, 1984, 1979, 1969) and Sebastiano Tusa (1979), are both representatives of the Italian Archaeological Mission in the Swat Valley, and have directed excavations at Aligrama, and Professor Stacul at Bir-kot-ghundai, Ghaligai, Kalakoderay and Loebanr III. Both have considerable knowledge of a range of sites in Swat that relate to the period under study here, in addition to later sites, and have attempted to synthesise this information into models that explain aspects of social organisation in this region. Dr Robin Coningham (Ali et al. 1998) of the Department of Archaeological Sciences, University of Bradford, has directed the most recent excavations at the Bala Hisar of Charsadda, and presented a model which aims to supersede the traditional geographical boundaries within which interpretation in this area has been carried out. Sir Mortimer Wheeler was the Director General of Archaeology in India from 1944 to 1947 (1976, 9-10), and as discussed in Chapter 1, his culture-historical legacy of interpretation is still influential in Pakistan today. Wheeler directed excavations at the Bala Hisar of Charsadda in 1958 (1962).

9.1 Extant theories and models

Tusa (1979, 675) used the results of excavations at Aligrama to develop a theory of cultural development in the Northern Valleys based on the inaccessible nature of the geography of the region. These factors, according to Tusa, designated this area marginal, with the nature of the terrain isolating the inhabitants. On the basis of pottery evidence from other Swat sites such as Ghaligai and Bir-kot-ghundai, Tusa notes that influences from both the Kashmir Neolithic and the Indus Civilisation are visible up to Swat PIV (c 1700-1400 BC) (ibid., 679). However, from PIV onwards, he claims that the material remains clearly show a contraction in outside contact and influence, and an internal development which led to the Gandharan Grave Culture (ibid., 690). Aligrama, which was first settled during PIV, has structural remains of above-ground dwellings only, being rectangular rooms of river cobbles and beaten earth floors, all of similar size, construction technique, and with similar artefacts recovered inside them (ibid., 685). This has led Tusa to speculate about the economic base at Aligrama, of which he says "one gets the impression of a farming-pastoral production pattern which is highly fractionated among the various small-size productive units probably coinciding with a family-type structure" (ibid., 687). In combination with the isolation of Swat, Tusa suggests that from PIV onwards in terms of political organisation, this region was not ruled by a hierarchical state structure, but was based on a type of federation (ibid., 692).

PIV is crucial in Tusa's model, and it is during this period that there is a move from temporary to permanent settlement structures, and this is interpreted as a move from presumed mobility or transience, to extended sedentary settlement and extensive pastoral activity (ibid., 690-1). However, the only direct reference to plant or animal exploitation by Tusa is made in association with the collection of miniature 'vases' which dominate the pottery assemblage. Apparently, these miniature pots "were used as containers for coagulating substances used in the treatment of milk and its by-products, as has been found in certain prehistoric and present-day societies" (ibid., 687).

Stacul has attempted to trace the causes of what he has interpreted as evidence for continuity and discontinuity between the different Swat periods (1969, 85), and one of the main identified areas of influence in the cultural development of the Swat sites is that of Northern China and the Himalayas (1996, 437). The inclusion of Swat PIV in an 'Inner Asian Neolithic' complex has been the basis of much of Stacul's interpretation and model of development (1996, 1994a,b). Based on the presence of artefacts such holed sickles (recovered only from Kalako-deray), and structural evidence such as pit-dwellings (at Loebanr III and Kalako-deray, but not Aligrama or Ghaligai), Stacul has drawn parallels between Swat and the Neolithic Burzahom in Kashmir, and from there, up into Northern China (1996, 437; 1994a, 708; 1994b, 235).

However, Stacul has also recognised that there are differences not only between the subsistence base of sites such as Burzahom, which has a mainly wild animal and plant assemblage (1994a, 710), but that there are many differences between the Swat sites. These differences are evident in structural and artefactual material (1994b, 238-9; 1996, 436), and also in terms of evidence for subsistence. Stacul discusses the heavy ground stone tool assemblage from Kalako-deray as different from that of any other Swat site. The number and range of tools has led Stacul to conclude that Kalako-deray was a specialised farming site (1996, 436; 1994, 235), yet again the presence of rectangular holed sickles are interpreted as the influence of Inner Asia. The interpretation of the large pits at Kalako-deray and Loebanr III as dwelling pits (1994a, 707) is part of the cultural package from Inner Asia.

In contrast, Dani's (1992, 1968) model, first developed over thirty years ago and still being published in text books, is based on population movement as the main source of development, including outside invasion (1968, 55-8). He says that in Pakistan 'tribes' had abandoned pastoralism and adopted sedentary agriculture by the end of the Indus Civilisation, and that these groups then became known as the Gandharan Grave Culture or people, but remained separate from, and contributed nothing to the development of the later Early Historic Urban sites at Charsadda and Taxila (1992, 395-6).

Dani also argues that geographic and environmental factors largely shape both economic and cultural responses, and so it "is natural that the hill zones of Pakistan, …where the advantage of natural river inundation and the annual refertilization of the soil by silt deposits enabled man to develop settled intensive agricultural systems" (ibid., 395-6). Dani believes that the Swat sites of Loebanr III, Aligrama,

Bir-kot-ghundai, Balambat in Dir, and Hathial at Taxila are all settlement sites associated with Gandharan Grave people, assigning them to one homogeneous culture, despite noting differences such as the presence of 'pit-dwellings' only at Loebanr III (ibid., 403).

Coningham and Sutherland (1998) challenged Stacul's interpretation of the Period IV pits at sites such as Loebanr III and Kalako-deray as dwellings, and then developed a new model of social and economic organisation to explain the archaeology of Swat between 1700 - 1400 BC. Coningham and Sutherland point out that Wheeler interpreted British Iron Age pits at Maiden Castle as dwelling pits, with postholes around the perimeter, suggesting some form of exterior superstructure (ibid., 180). Recent work suggests that the pits were more likely to have been used for grain storage, as has been shown by Cunliffe at Danebury, on the basis of environmental evidence. At Danebury, Coningham and Sutherland note that underground pits were thought to have been used for seed grains and to hide goods from raiders, and above ground storage for grain intended for consumption (ibid., 181).

The Kashmir-Swat Neolithic has to date, been interpreted as a period of sedentary agriculture, which was made possible by pit dwellings in the harsh winters. Coningham and Sutherland suggest that the population was perhaps transhumant, and used underground pits to store seed grain seed during the winter, in an underground 'reducing atmosphere' (ibid., 183). This would also explain the lack of evidence for other dwellings because they suggest that transhumants would erect only lightweight, temporary buildings (ibid., 184).

Wheeler bases the appearance of early prehistoric urban forms in north west Pakistan on a culture historical - diffusionist model. He states that the evidence available from his excavations at Charsadda, and from Taxila and other comparable sites such as Begram in Afghanistan, shows that prior to Persian rule in this region none of these cities existed (1962, 5). This interpretation of the evidence in conjunction with his relative pottery chronology both support his founding date of c. third quarter of the 6th century BC for the Bala Hisar (ibid., 33). While Wheeler says that the Bala Hisar grew in an organic way, rather than being a planned city in the Greek sense of city planning (ibid., 12), it is clear that he believes the site to have always been urban, or civic in nature, and thus always to have displayed the urban characteristics discussed further below. He does not allow for any transitional phases, or even pre-urban forms to have been present at this site, indeed, his is a model of discontinuity in development in the Vale of Peshawar.

However, further work in the Southern Plains region since the 1960s has allowed significant re-interpretation of the possible early origins of the Early Historic cities here. On the basis of an analysis of pottery, including the early Red Burnished ware (Wheeler's soapy red ware) from both Taxila (Hathial) and Charsadda (Bala Hisar), Allchin was able to conclude that with regard to "Gandhara there was an unbroken occupation at Pushkalavati through from the early

Red Burnished ware period, which we have identified as originating prior to Achaemenid rule" (1995, 130). The new chronometric dates from the Bradford-Peshawar excavations have confirmed this hypothesis, and suggested regions of contact and influence other than the Achaemenid Empire to the west.

9.2 Urban archaeology and the Southern Plains

In order to determine whether faunal and floral assemblages from urban and rural sites can be characterised as such, some of the general definitions of 'urban' that relate to archaeology will be considered, and then applications of this theory to research in South Asia will be noted. Whether the material in this study can be used to distinguish between rural and urban will then be assessed in the general discussion and testing of the archaeological environmental data.

The work of Childe in urban theory relating to archaeology has provided some of the fundamental ideas still influential today, and is based on a series of progressions and elements, such as script, that could then be applied to any given site to determine whether it was urban (1957, 36-7). The conclusion of his study of the definition of urban gave rise to his trait list of: size; full time specialisation; surplus; monumental buildings; unequal distribution of social surplus; text; predictive sciences; naturalistic art; long distance trade, and non-kinship based residence (McNairn 1980, 99). The seminal nature of Childe's work is demonstrated in its use as a starting point for many other scholars in South Asia attempting to define urban sites and development, such as Coningham (1995, 56) and Erdosy (1995). Allchin (1993, 73) says that both the Bala Hisar and Hathial cover around 12 hectares each in area, but apart from size, other traits from Childe's list are not necessarily apparent at either the Bala Hisar or Hathial (Allchin 1995, 127) and only some such as full time specialisation and town planning (Marshall 1951, 20, 90-91) evident at the Bhir Mound at Taxila, and thus there is some question about its city status (ibid., 20).

Allchin (1995, 124-5) offers three possible causes for the development of cities, specifically the cities of the second, or iron, urban period in north western Pakistan. The first of these three causes is the indigenous growth of cities or a city, arising as the central focus of a smaller group of settlements. The second is the emergence of cities as a result of colonisation, or colonial expansion, whereby colonists either evolve new cities or take over and develop existing ones. Third, cities may be the result of Imperial expansion, such as the Achaemenid conquest and settlement in the north west. Environmental change is also suggested as an explanation for different urban development in the regions of the Northwest Frontier and Punjab, and Sindh, with Sindh becoming drier than the north in the post-Harappan period (ibid., 125).

Shaffer, in his analysis of the Harappan 'socio-cultural' complex, concludes that it grew quickly and spread quickly over a large geographical area. He suggests that this might be the product of the cattle worship and keeping, requiring room for expansion (1993, 62). He attributes the so-called the second or Early Historic urban period to a shift from pastoralism and agro-pastoralism to more intensive agriculture, although cattle in particular he argues retained an ideological importance testified by their artistic representations, in addition to still having an economic value (ibid., 63).

In terms of diet and subsistence evidence from large urban sites, Weber (1999) and Patel (1997) were able to propose certain features of the Mature and Late Harappan that indicated a difference between urban and rural sites (see Chapters 6 and 7). These differences appear to be subtle, and often dependent on the presence of both urban and rural sites of comparable dates to enable such discrimination. Insoll, in his discussion of Islamic diet suggests that distinguishing between Islamic and non-Islamic archaeological faunal assemblages is very difficult, and also based on very small or subtle differences (1999, 95). Work at the Jordanian site of Pella suggested little change with regard to sheep, goat and cattle in faunal assemblages from Christian to Muslim periods, but the introduction of camel has been recognised here as a possible 'marker' of Islam (ibid., 98-9).

9.3 The subsistence strategies revealed in the ethnographic material

The ethnographic work outlined in Chapter 8 clearly shows that subsistence patterns within the study area are complex and multiple, and no single model outlined above could account for the multiplicity of approaches to exploiting the plant and animal base in the two contrasting regions. There are at least four different variants of seasonal transhumance practised in and from Dir and Swat in addition to year-round sedentary agro-pastoralism. In the Vale of Peshawar where the fundamental approach to subsistence is dominated by sedentary agriculture, this region is open to a number of transhumant groups during at least part of the year, interacting on a regular basis with permanent residents. These multiple strategies are shown in Figure 4.

The mobile groups have herds of sheep and goat, with cattle, donkey and some buffalo, as do the sedentary groups of Swat and Dir. Both reported growing wheat, maize, potatoes, vegetables, and sometimes rice: that is both summer and winter crops. In the Charsadda District, the range of animals kept and crops grown are much narrower, with cattle and buffalo dominating the domestic animals kept, with each farm keeping far fewer animals than their mobile counterparts. The crops grown in Charsadda for human consumption are wheat and sugar cane, the latter as a cash crop. The only other significant crop was the clover type, also a cash crop, grown for animal food. It could be argued that as the residents of Charsadda have access to a range of plant and animal products through trade and have a source of wealth in their cash crops, they do not need to diversify, whereas the mobile groups not only are committed to herding and maintaining animals as a source of wealth, but their mobility and access to grazing and suitable land for agriculture, allows them to produce a greater variety of animal and plant foods.

As the transhumant groups are all seasonal, then some type of seasonal patterning in material remains of an environmental nature could be expected, in particular through age at death of animals, which may relate to time and place of birth and place of death. Plant seasonality could also be important, and information from weed assemblages associated with crops could be of great value in determining the time of harvest. Recovery and identification of elements such as chaff could help to distinguish between crops grown locally, or those stored and later traded or consumed in a different area. In terms of crops, the growth of rice, a summer crop, in the Northern Valleys, strongly suggests a population present during the growing season to tend the crop. While rice is commonly grown in Swat and Dir today, it is not so common in the Vale of Peshawar.

The results of the interviews indicated that while sheep and goats tended to give birth around April and May (spring), this could occur either while at the winter residence in the case of winter transhumants or back at home in the valleys, or prior to moving within the valleys in the case of the inter-valley or summer transhumants. Slaughter of young animals was reported as uncommon, and so the appearance of juveniles in an animal bone assemblage could be considered unusual. Indeed, as Gilbert (1983, 109) noted in relation to Iranian herders, the practice of separating animals giving birth from the main herd, possibly at some distance, can result in a distortion of the animal bone assemblage.

The domination of the modern Charsadda animals by cattle and buffalo, with their primary use for traction and dairy products, means that they are unlikely to be slaughtered before passing their physical peak. This means that age at death within faunal assemblages may not reveal information about subsistence strategies relating to mobility. Rather, more is likely to be learnt from the relative proportions of the animals present. The physical requirements of such animals and buffalo, and their unsuitability for steep terrain, means that they are unlikely ever to dominate animals kept in the Northern Valleys. Further, they are not as suited to mobility as goats and sheep, and even cattle are, hence the relatively recent acquisition of buffalo by transhumant groups. Conditions in an area such as Charsadda District are far better suited to buffalo management, while the time required to ensure that flocks of sheep and goat have access to enough grazing land does not readily fit in with the demands of growing sugar cane and wheat reported by the Charsadda farmers. However, for the mobile groups who have large flocks, and require year-round access to a certain quality of grazing, such an investment is central to the pastoral element of their economy.

In summary, the ethnographic work showed that there are many responses to living in the Northern Valleys of Swat and Dir, involving a combination of seasonal movement, sedentism, farming, animal husbandry. While it is clear that the ethnic affiliations of those involved can shape the approach taken to subsistence, this is not fixed, and other factors such as the area of permanent residence and its climate and geography must also be taken into account.

9.4 Testing the new model with the archaeological material

To test the new model developed from the ethnographic interviews and observations, with the archaeological material, the main trends and results evident in the analyses of environmental material from both regions are first summarised. How well the existing models deal with this and other archaeological information relevant to economic organisation will be considered, and this will be followed by a discussion of the new model and how well the archaeological data fit the model.

Northern Valleys: a summary of the evidence.

1. Within the Northern Valleys there is evidence for the use of both wild and domesticated plant foods during the same period.

2. There is evidence for a range of cereals, including wheat, barley and rice, which is also evidence for both summer and winter crops in this region.

3. The presence of barley alongside wheat could be interpreted as production of the former for animal food, while wheat is more suited to human consumption.

4. The animal assemblage is dominated by sheep/goat and cattle, with NISP counts showing cattle the most numerous at all the sites, while MNI estimates show sheep/goat as the most abundant, with the exception of one site.

5. The MNI estimate shows that cattle dominated the assemblage at Bir-kot-ghundai, the urban site, yet the difference between cattle and sheep/goat MNI estimates at Kalako-deray, the least accessible and most northern of the sites discussed is minimal.

6. The differences between the environmental material from the sites is as great here as between these sites and those of the Southern Plains.

How do these conclusions fit with the existing models? The existing models discussed above either suggest a move from pastoral activity to sedentary agriculture (Dani 1992; Stacul 1994a,b, 1996; Tusa 1979), or propose seasonal transhumance (Coningham & Sutherland 1998). The significance of the animal assemblages, and the presence of barley suggest that pastoralism was still an important element, if not a major aspect of the Northern Valleys subsistence base. The nature and seasonality of the crops also strongly suggests that agriculture was a vital part of subsistence during the period of study. The model of transhumance, developed as an alternative to the interpretation of the large pits as winter dwellings, is important as it offers alternatives of economic organisation, and so opens up many possibilities. However, winter transhumance alone is not sufficient to interpret the environmental material. The complexity of the archaeological and environmental material at the different Swat sites needs a model with multiple layers rather than offering transhumance versus sedentary, either/or options.

Bala Hisar of Charsadda: a summary of the evidence.

1. Only domesticated types were identified in the plant assemblage.

2. Wild animals were represented by a small amount of bone, and this may suggest that hunting was occurring around the site, or alternatively that wild, hunted remains were being brought to the Bala Hisar from other regions.

3. The presence of buffalo is very significant in terms of environmental factors and economic organisation, as the requirements for buffalo indicate.

4. There is a very narrow range of types in both the plant and animal assemblages, which may suggest that the site occupants or consumers were removed from the process and the areas of production.

5. Wheat was the only cereal recovered and identified, with the exception of small quantities of rice husk recovered from later contexts (Phase C).

6. Strictly speaking, lentil and wheat are both winter crops, meaning no summer crops (except the rice) were noted here, though it is possible in the Vale of Peshawar to have two wheat crops per year with the use of irrigation.

Urban Bala Hisar does have similarities with material from urban Harappan sites, and in particular the presence of buffalo in relatively large quantities in terms of the total animal bone assemblage could be interpreted as important with regard to economic organisation (Patel 1997, 111). The Bala Hisar plant and animal bone assemblages differ from Northern Valley sites most significantly on the basis of the presence of buffalo, and that wheat is the only cereal recovered and identified, with the exception of the small quantity of rice husk from later contexts.

9.5 The archaeology and toward a new model

Within the Northern Valleys there are similarities between the sites on the basis of pottery, and other small finds, such as terracotta figurines; and differences, such as the distinctive ground stone tool assemblage from Kalako-deray, which is not replicated at any other site (Stacul 1994b, 235). There are also differences in terms of the structural sequences. There are the large dwelling pits at Loebanr III and Kalako-deray in the early phases of PIV (Stacul 1996, 435). At other sites such as Aligrama and Bir-kot-ghundai, there are pits, but these are not interpreted as dwellings, while at Ghaligai there were no pits recorded (Stacul 1987, 63). The sites range in size from Bir-kot-ghundai with its later (c. 3rd century BC) fortification wall (Callieri 1992, 342-3) which is interpreted as a town at this stage, and may perhaps be considered an incipient urban site during PIV, to Kalako-deray, covering some 2000 square metres (Stacul 1993, 69). The positions of the sites vary. For example, Bir-kot-ghundai is located on a hill-slope in the lower main valley at its widest point, thus giving easy access to land suited to grazing and agriculture. By comparison Kalako-deray is situated in a side valley, on a single hill top, at some height from the nearest stream. The Kalako-deray hill top is in an area of hills, and there is little in the way of flat, open ground around it. From the Northern Valley sites there is evidence for "Multi-directional and long-distance contacts" (Stacul 1980, 76) in the form of pottery and tool types, that have been interpreted as showing contact with and influence from the Indus Valley, Northern China and Iran.

When the archaeological environmental data are fitted against the ethnographic data, it is clear that in both cases there is no single subsistence strategy emerging for either area. There is in fact a mosaic of subsistence strategies existing side by side in both areas, allowing for contact, and the exploitation of what Barth (1956, 1079) calls different ecologic niches, and the separation of different ethnic groups on the basis of their chosen strategy. The ethnographic model shows the complexity and the overlap in terms of the types of animals and crops, and geographical area of each group. The archaeological data would fit well in a model where crops and animals are both important, though different species and seasons may be important to different groups. However, the overlap between these groups at all stages, and between both sedentary groups in the Vale of Peshawar and mobile groups from the Northern Valleys means that distinguishing between the settled and the mobile groups at any given site would be dependent on determining more subtle discriminants than is possible with the current data sets.

Contact between the Northern Valleys and a wide range of areas is fundamental to this new model. Stacul has proposed his model of the 'Inner Asian Circle', which is based around a largely enclosed, self-referencing area broadly encompassing the modern areas of NWFP, Kashmir and the bordering areas of China. Tusa argued for the isolation of the even smaller geographical area of Swat alone, as did Dani. However, the new model which proposes regular seasonal mobility, with groups moving outward from the Northern Valleys in directions regulated by season and herd requirements, suggests complexity in both subsistence strategies and contact. A stylised pattern of contact in the Northern Valleys during PIV is shown in Figure 8, which takes into account trade routes and seasonal movements for grazing. In addition to the mobility of the transhumant groups, the aspect of trade is one which has been noted by Stacul (1996, 437), Dani (1992, 417) and Tusa (1979, 681). However, their interpretations of the significance of trade is used to support their various theories of cultural development in the Northern Valleys. For example, Tusa suggests that trade with regions outside the Northern Valleys themselves is limited and confined largely to small amounts of high quality objects (1979, 691), while Stacul emphasises the links with Kashmir and Central Asia, through goods such as the jade objects (1996, 437; 1984, 209-10).

9.6 Occupation dates from the Northern Valleys and the Bala Hisar of Charsadda

Taking into account the survey work (Ali 1994; Dani 1968; Facenna 1965; Khan 1993; Rehman 1968), the relative dating schemes (Dani 1992, 1968; Stacul 1969; Wheeler 1962) and the radiocarbon estimates (Ali et al. 1998; Coningham pers. comm., 2000; Possehl 1994; Stacul 1987), it is possible not only to begin to develop a chronology for both of the study regions that shows an overlap in occupation (see Chapters 4 and 5), but also to consider settlement patterns in terms of the possible influence of older and newer settlements. Given the current data, it seems clear that occupation in the Northern Valleys for the Late Bronze and Early Iron Age period is of greater recorded antiquity than that of the Vale of Peshawar, in particular the Charsadda District. The reasons for this, and the effects of the growth of the site of the Bala Hisar are likely to be many, and complex. However, some of them will be considered here, both as a refutation of the standard models, and as theories arising from the new model, which need testing with archaeological data.

9.7 Implications of the new model: further discussion

Wheeler wrote a lyrical description of the Vale of Peshawar in his introduction to the excavation report of Charsadda, describing the fertility of the area, and making the assumption that such fertility had its roots in prehistory, which would undoubtedly have been one of the main reasons why this area was settled (1962, 1). Reports in the 1897-98 Gazetteer (Punjab Government 1898, 24-5) are generally united in describing the Vale of Peshawar as an area of mixed cultivation and wasteland. Other early reports suggest that this area was not always as green and lush as Wheeler described it. For example, McMahon and Ramsay (1901), Political Agent and Assistant Political Agent respectively for Dir, Swat and Chitral, compiled information about this whole area for official reports at the end of the 19th and beginning of the 20th century. They described a district called Sam Ranizai stretching from the range of hills to the south of Swat (ie the Malakand Hills), down into the plains to the south (ie the Vale of Peshawar) (ibid., 12). They said the area was dry and dependent on irregular rainfall, resulting in frequent crop failure, and only when the residents had dug wells to provide irrigation could they depend on good crops (ibid., 13).

This raises questions about the role of irrigation in this region. In Chapter 3 the effects of the major modern irrigation schemes such as the Upper and Lower Swat Canals were briefly discussed, and the significance of the major rivers in terms of irrigation noted. Little work has been carried out to date on the archaeology of irrigation in this region, but it is reasonable to suggest that the intensity of settlement in the Vale of Peshawar today is in part a product of the private and government irrigation schemes. Therefore, it is also possible to suggest that prior to intensive irrigation, settlement would have been more restricted. Settlement survey in this region (Ali 1994) clearly indicates that intensity of settlement in this region was achieved only during the Buddhist period, and while there are of course a number of methodological and site preservation issues that affect surveys of this nature, it may be possible to argue that only with the Buddhist sites came the concurrent exploitation of the surrounding land. Further, little is known about prehistoric tree or bush cover in this region. Greater cover would have required clearance prior to intensive use of the land.

In contrast to the Vale of Peshawar, the climate and geography of Dir and Swat have given rise to a region with greater natural fertility, although not without its own difficulties in occupation. Irrigation from both rainfall and khwars is well documented, both in modern and historical sources, and the range of plant remains from the Northern Valleys sites certainly appears to confirm that the occupants were taking advantage of the fertility of this area. The surveys of Swat, and to a lesser extent Dir (eg Dani 1968; Facenna 1965; Stacul 1987), have demonstrated that there is a high intensity of pre-Buddhist sites, certainly greater than in the Vale of Peshawar for the corresponding period.

Evidence for long distance contact at the Northern Valley sites for the periods prior to 1400 BC, (the earliest date given thus far for occupation at the Bala Hisar, and also the estimated end of Swat Period IV by Stacul (1987)), include pottery similarities between the Ghaligai and Bir-kot-ghundai assemblages and Harappan material (Stacul 1996, 438; 1984, 205, 209). In Chapter 3, the presence and antiquity of many of the major routes through this region were discussed, and the importance of such routes to the cultural development and occupation sequence has been acknowledged (Stein 1921, 35), although it has also been noted that Charsadda is not situated on "an obvious crossroads of major routes like Taxila" (Allchin 1995, 15). Indeed, trade and influence from other areas, suggested by pottery styles, have been recognised from a number of sources at many of the Northern Valleys sites (Stacul 1994a,b; 1996).

Therefore, it could be argued that the Northern Valleys prior to c.1400 BC were a region of great significance. Contrary to the models of fluctuating isolation and cultural marginality, the combination of fertility and the means of exploiting this through seasonally mobile subsistence strategies, and the pivotal nature of the region in terms of trade routes, suggest that this region was one of strategic and cultural importance. Stacul's arguments for inclusion within an 'Inner Asian Circle' can be refuted in part by pointing out influences from other areas, such as the Harappan style pottery at Bir-kot-ghundai, and other links to the south (Stacul 1980, 61-2). The interpretation of the large pits as pit dwellings, one of his major arguments for linking Loebanr III to Burzahom, is challenged by the proposal of transhumant occupation of the site as one solution to surviving harsh winters (Coningham & Sutherland 1998). While there are other important similarities between Swat and the Kashmir Neolithic, there are also as many similarities in terms of artefacts with other regions that are connected to Swat by trade routes to the north west, in particular Iran, and also to the south.

Both Stacul and Dani argue for discontinuity evident in the archaeological record, and give as the explanation for this, incoming groups, ie invasion or migration theories. This is based on changes in pottery styles (Stacul 1996, 1987), and changes in burial styles within the Gandharan Grave Culture (Dani 1992, 52-5). However, Stacul (1996) also argues for internal development between other phases and periods in Swat, with continuity and development of pottery forms. An alternative model could be postulated, one based on both extensive long distance trade (on the basis of exotic goods on site) and regular mobility of a section of the population (on the basis of the ethnographic data, and similarities between the predicted ethnographic model and the archaeological material). Figure 8 is a simplified version of both trade and subsistence movement in the study area. This model would then explain change, even major cultural change, as an amalgam of external influence diffused through trade, and internal development as a result of such stimulus. Contact between Valley and Plains is further indicated by the presence of rippled rim type pottery in the lowest levels of the Bala Hisar of Charsadda (Wheeler 1962, 25, 28) and in PIV layers at Loebanr III (Stacul 1977, 245) and from Timargarha 1 (Dani 1967, 249-50).

This suggests that the Northern Valleys, or perhaps more specifically the Lower reaches of both Swat and Dir, may have been the settlement focus of this whole area in the Later Bronze Age period of prehistory. This is supported by the archaeological evidence to date such as the sites known through survey and excavation, and also by a model of combined mobile pastoralism and sedentary agriculture, which takes advantage of the seasonal environmental changes, and cements ideas of contact and influence into the indigenous development of this area. If the fertility of the area is accepted on the basis of the environmental data, and population expansion up to PIV (1400 BC) on the basis of the archaeological structural data (Stacul 1984, 209; Tusa 1979, 681), then it is possible to argue that occupation development in the Vale of Peshawar may be result of increasing population and attendant pressures in the Northern Valleys. While the Northern Valleys demonstrably have very fertile land, this is also limited in area, and so a population increase over an optimum number would require new strategies.

Transhumance in this region is not necessarily suggested as a response to this pressure (eg Lees & Bates 1974, 189-90) as the environmental material could be interpreted as showing that it is already in place by this time, but rather transhumance is seen as a means of contact with new areas that may have been considered suitable for expansion. Rice agriculture may be an important factor in the success of the Northern Valleys. Rice is a summer crop, and one that requires a degree of co-operation to arrange and maintain irrigation, and it is entirely compatible with the practice of seasonal winter transhumance. Figure 7 shows a three phase development within the Northern Valleys and the Plains. This illustrates the potential contact between the two regions, and the subsequent shift in the importance of each.

The Bala Hisar of Charsadda is recognised as a major city of this region during the Iron Age, or Early Historic period (Ali et al. 1998, 1; Allchin 1995, 15-16). This development must have relied on agricultural and pastoral exploitation leading to surplus, perhaps beginning with a gradual transfer of dependence from the Northern Valleys to the surrounding Vale of Peshawar as the ability to manipulate this landscape grew. Although as noted above, Charsadda does not lie on any major trade routes, it is probable that trade played a role in its growth (see Figure 8 also). Therefore, it is suggested that while the focus of trade and settlement was firmly based in the Northern Valleys, where the fertility of the land and the multiple subsistence strategies allowed population expansion and cultural development and change, the Vale of Peshawar was a secondary area, useful perhaps for winter grazing for the winter transhumant group. However, the population pressure in the Northern Valleys meant that expansion south was a logical progression, resulting in emphasis on sites such as the Bala Hisar as a focus for trade and mobile groups during the winter. This shift in focus, especially in terms of trade, may have had a long term, far-reaching effect on the viability of the Northern Valley sites. As the Bala Hisar grew, the Northern Valley sites appear to have contracted, until the Buddhist period, where archaeological surveys and excavation, and historical records, show that the landscape of the Northern Valleys was dominated by Buddhist sites with accompanying agricultural exploitation.

9.8 Summary and conclusions

The two regions explored in terms of their subsistence strategies can be described as both pastoral and agricultural on the basis of the environmental evidence, rather than being exclusively one or the other, or changing from one to the other over time. It also appears that influences in terms of material culture have been derived from a number of sources, and not a single area, such as the 'Inner Asian' region. The practice of transhumance has been demonstrated as both an environmental and a cultural response (Geddes 1983, 58; Kuzmina 1997, 286; Lees & Bates 1974, 189-90), and it allows contact with many areas outside the immediate geographical boundaries. The tentative Harappan style links, noted in pottery from Birkot-ghundai and other Swat sites is perhaps indicative of such contact and influence. Mughal's (1994) study in the Cholistan Desert identified a significant nomadic pastoral group throughout the Harappan, and Possehl (1979, 546-8) has also suggested pastoral nomadism as an adaptation throughout this time period. Movement by those designated Harappan, as well as by those belonging to the Northern Valleys suggests possibilities for contact and exchange of goods and ideas.

The archaeological environmental data can further be interpreted and re-interpreted with respect to the new ethnographic model, with two trends or conclusions evident. First, that rural and mobile means a greater range of crops, greater exploitation of range of animals, as shown at the site of Loebanr III, and second, sedentary urban suggests contraction in range of both plants and animals, as shown at the Bala Hisar of Charsadda. These trends are likely to be the result of many factors, including access.

The radiometric dates from the archaeological sites (Table 9.1) show that contact between the two regions was quite possible, and this is supported by the artefactual data. That the period under discussion is one of incipient urbanism at the Bala Hisar, while the sites in the Northern Valleys appear to be already well established, if developing culturally, has interesting implications for the nature of this contact. It was suggested above, that during PIV of the Swat chronology, the Northern Valley sites may have dominated this region in terms of trade and position on the trade routes to Central and Western Asia through to South Asia, situated as they were in a fertile region, with a range of subsistence strategies designed to exploit the region, and two crops per year to take maximum advantage of the land. If one of these strategies included long distance seasonal transhumance, movement during winter would most sensibly have been south, thus to the Vale of Peshawar. As noted above, without irrigation, the fertility of this region is not necessarily reliable, and these mobile groups may well have been aware of this, and perhaps the potential of this region. Thus it may have been pressure from the north that influenced the development of the incipient urban form here, perhaps with the summer crop of rice playing a pivotal role in feeding both areas.

The analysis of the ethnographic information presented in Chapter 8 demonstrates that there are many approaches to subsistence in the Northern Valleys of NWFP, while on the plains to the south, there is a far narrower range of strategies. The difficulties of using modern information for interpreting archaeological data have been discussed at some length in Chapters 2 and 3, and it is recognised that many factors in modern land use may have altered these subsistence strategies. However, it is also possible to see strong similarities between the ethnographic model and interpretations of the archaeological environmental data. To test the model further would require more data, both archaeological environmental data and detailed ethnographic information from interview and observation. Suggestions for possible further work are outlined in greater detail in the next chapter.

Table 9.1 Summary Chronology: ^{14}C Dates from the Bala Hisar of Charsadda & the Northern Valley Sites

Bala Hisa		Northern Valley Sites		
Ali et al		Swat Chronology		
		I	Ghaligai	2970-2920 BC
				2520-2230 BC
		II	Ghaligai	2180 BC
				1980-1870 BC
		III	Ghaligai	1950-1920 BC
				1660-1560 BC
Ch.VIII		IV	Aligrama	1360-1300 BC
Phase A	1420-1140 BC			1710-1690 BC
Phase B	1380-1090 BC			1210-1090 BC
			Loebanr III	1730-1600 BC
				1560-1225 BC
			Timargarha	1590-1470 BC
Phase C	1260-990 BC	V	Aligrama	1540-665 BC
	800-250 BC		Timargarha	1000-800 BC
Ch.VI				
Phase C	770-410 BC			

sources: Ali et al 1998, Coningham pers. comm. 2000, Possehl 1994, Stacul 1987

Chapter 10
Agriculture and Pastoralism in the Late Bronze and Iron Age, North West Frontier Province, Pakistan

The aim of this research was to explore subsistence strategies in two contrasting areas of NWFP during the Late Bronze and Early Iron Ages, and to develop a new explanatory and predictive model to relate conclusions from the archaeological environmental assemblages to economic organisation. This aim has been achieved through the analysis and interpretation of both new and published archaeological environmental data, which has then been used to test a new model developed from ethnographic observation and interview. This study has presented the results of integrated environmental analysis, and demonstrated that the archaeological material can be interpreted as evidence for an integrated subsistence system.

The environmental assemblages from all the sites are small, and in some cases (eg Timargarha 1) have been of little value beyond confirming the presence of certain animal types in the region during the study period. However, it has also been demonstrated that many of the established models and theories with aspects relating to subsistence practice in both study regions were developed on the basis of such small samples, and there has been little attempt to interpret either region as a coherent whole, or to consider the two study regions in relation to each other. By considering each site as both separate entities, and also as part of a wider region and then area, it has been possible to put forward an entirely new interpretation of the ways in which plant and animal foods were exploited during the later Bronze and Early Iron Ages, or the mid second to mid first millennia BC, in the Northern Valleys and Southern Plains of NWFP, and how such exploitation might relate to the economic organisation of the site occupants.

The archaeological data show that a greater range of plants, including both summer and winter harvested crops, were being produced in the Northern Valleys than on the Plains, while sheep/goat and cattle were the main domesticates present on all the sites. However, within the five Swat sites there was greater variation in such things as the relative proportions of the animals, the presence and quantity of wild plants and animals, and the recovery of artefacts interpreted as associated with plant processing and animal activity such as grind stones, hoes, and terracotta figurines. Indeed, there was shown to be greater variation between the sites in the Northern Valleys themselves, than between these sites and the Bala Hisar of Charsadda in the Vale of Peshawar in certain respects. However, there were also important differences recognised between the two regions in terms of the environmental archaeological material. One of these differences was that buffalo was present at the Bala Hisar from contexts in the earliest occupation phase, while it was recovered only from Aligrama, and only from the PV levels here. Another difference between the Northern Valley sites and the Bala Hisar was that there was a much narrower range of plant foods recovered from the Bala Hisar, with wheat virtually the only cereal recovered. The presence of buffalo in the plains, but more limited in the hills can be argued as the product of more suitable environmental conditions in the former area, and may also be related to sedentary agricultural activity. The limited range of plant foods, supplemented only in later contexts by rice, may be the result of both limitations imposed by lack of irrigation, ie environmental conditions, and the constraints imposed by a developing urban area which is cut off from the surrounding farm land.

The introduction of the new ethnographic material is very important for a number of reasons. The existing theories and models all suggest one extant subsistence strategy at any given period in the prehistory of this area, and so one model of economic and social organisation. Despite the fact that these models are often contradictory, and in certain examples only loosely based on the actual recovered environmental data, they all propose that in both the Northern Valleys and in the Southern Plains, at any given time during the one thousand or so years covered by this research, that there was one dominant mode of subsistence. Pastoral nomadism is generally considered the earlier adaptation, followed by sedentary agriculture as the area developed (Dani 1992; Stacul 1987), with interpretations of the archaeology supporting this. Further, no attempt, whether environmental or otherwise, has been made to distinguish possible mobile groups in archaeological assemblages.

The exception to this is, of course, the work by Coningham and Sutherland (1998) which offers the possibility of regular seasonal transhumance as a means of explaining the numerous large pits of Loebanr III and other sites mainly the numerous large pit, without interpreting them as winter pit dwellings (Stacul 1987). However, Coningham and Sutherland propose a single alternative interpretation in the form of seasonal transhumance. Yet the ethnographic interviews were very clear in demonstrating that there is no single clear pattern in modern subsistence practice. There is a range of approaches to obtaining a living from the land, and even within the definition of transhumance, there are at least four very different types of seasonal movement centred round (though perhaps periodically moving beyond) the Northern Valleys. These transhumant groups regularly have contact with year round sedentary groups, both in the Northern Valleys and in the Southern Plains, and in the case of the winter-transhumant group, reports were made of returning to the same village or area over a period of many years and spending up to six months of every year there. The new model based on the ethnographic work emphasises integration. It examines integration of mobile and sedentary strategies, with contact and interaction between the two strategic approaches, as well as integration of both plant and animal assemblages.

This research has been of great significance in demonstrating the value of the analysis and interpretation of environmental material. While the limitations resulting from sampling and recovery programs and the size of the assemblages have been acknowledged, and indeed have restricted the subsequent work, what has been achieved with the material available has allowed a new approach to understanding more about the subsistence strategies in the study area. In a region where environmental archaeology is not a high priority, it is hoped that presenting results based on a combination of environmental data and other archaeological material will encourage future archaeological research to include more information about systematically collected plant and animal assemblages.

Ideally, archaeological environmental specialists should be involved in any project from the initial research design onward. If this were to happen in NWFP then it would be possible to sample sites effectively, and begin to build a corpus of plant and animal bone data that could be used for more sophisticated analyses and so address more complex questions, than has been possible here. Further ethnographic work directed expressly at understanding subsistence practice would also be of value in providing an interpretative framework for archaeological data. With regard to understanding more about mobile strategies and their material remains in this region several approaches are possible.

First, fieldwork aimed at developing a methodology for detecting transhumant seasonal sites and temporary nomadic sites using both environmental and structural data would be of fundamental importance. While such recording work has been carried out Western Asia with pastoral nomadic groups (eg Cribb 1991), there has been little similar work in NWFP. Given the seasonal nature of many of the subsistence groups already identified through the ethnographic work, it would be necessary to conduct fieldwork to coincide with the full range of activities.

Second, an exploration of the problems of transferring ethnographic inference to archaeological material could be carried out in conjunction with such work. In Chapters 2 and 3 some of the problems of using modern data and models with regard to interpreting archaeological data were discussed. It seems that many researchers now feel that finding corresponding patterns in data is ample justification for such interpretations, and while this may well be true, this is surely an area that would benefit from greater methodological rigour, particularly in South Asian archaeology.

Applying the developed methodology to selected archaeological data sets from Pakistan and perhaps other areas in South Asia would allow new interpretations aimed at understanding variation in subsistence strategies. While much is assumed about the antiquity of many subsistence practices in South Asia, it is necessary to back up these assumptions and assertions with hypotheses and models that can then be tested with data. In his description of the Bala Hisar and its setting at the time of his 1958 excavation there, Wheeler said that it comprised "an area nearly 2 miles square...massively piled with vestiges of habitation amidst well-watered farmlands where the seasonal routine includes, and must always have included, the arrival of picturesque nomads from beyond the mountains in November, and their not-unwelcome departure in the spring" (1962, 1). While this is a very perceptive comment, its lack of support in terms of data points towards one of the major problems in South Asian archaeology.

Bibliography

Ahmed, A.S. 1991. *Resistance and Control in Pakistan*. London & New York: Routledge.

Ali, I. n.d. *Excavations at Hund on the banks of the Indus: the last capital city of Gandhara* (unpublished excavation report).

Ali, T., Coningham, R., Durrani, M.A., & Khan, G.R. 1998. Preliminary report of two seasons of archaeological investigation at the Bala Hisar of Charsadda, NWFP, Pakistan, *Ancient Pakistan* XII: 1-35.

Ali, I. 1994. Settlement History of Charsadda District, *Ancient Pakistan* IX:1-164.

Allchin, B. 1985. Ethnoarchaeology in South Asia, in J. Schotsmans & M. Taddei (ed.), *South Asian Archaeology 1983*: 21-33. Naples: Naples Istitutio Unversitario Orientale.

Allchin, B. 1986. The ground, pecked and polished or heavy stone artefacts, in F.R. Allchin, B. Allchin, F.A. Durrani & M.F. Khan (ed.), *Lewan and the Bannu Basin. Excavation and survey of sites and environments in North West Pakistan*: 41-64. Oxford: BAR International 310.

Allchin, B. 1994. South Asia's living past, in B. Allchin (ed.), *Living Traditions. Studies in the Ethnoarchaeology of South Asia*: 1-11. New Delhi: Oxford & IBH Publishing Co PVT. Ltd.

Allchin, F.R. 1993. The urban position of Taxila and its place in Northwest India-Pakistan, in H. Spodek & D.M. Srinivasan (ed.), *Urban Form and Meaning in South Asia: The Shaping of Cities from Prehistoric to Precolonial Times*: 69-81. Washington: National Gallery of Art.

Allchin, F.R. 1995. *The Archaeology of Early Historic South Asia*. Cambridge: Cambridge University Press.

Allchin, R. & Allchin, B. 1997. *Origins of a Civilization* New Delhi: Viking.

Allchin, F.R., Allchin, B., Durrani, F.A. & Khan, M.F. (ed.) 1986. *Lewan and the Bannu Basin. Excavation and survey of sites and environments in North West Pakistan*. Oxford: BAR International Series 310.

Barnard, A. & Spencer, J. 1996. *Encyclopedia of Social and Cultural Anthropology*. London & New York: Routledge.

Barth, F. 1969. The system of social stratification in Swat, North Pakistan, in E.R. Leach (ed.), *Aspects of Caste in South India, Ceylon and North-West Pakistan*: 113-146. Cambridge: Cambridge University Press.

Barth, F. 1956. Ecologic relationships of ethnic groups in Swat, North Pakistan, *American Anthropologist* 58: 1079-1089.

Belcher, W.R. 1991. Fish Resources in an Early Urban Context at Harappa, in R.H. Meadow (ed.), *Harappa Excavations 1986-1990. A Multidisciplinary Approach to Third Millennium Urbanism*: 107-120. Wisconsin: Prehistory Press.

Belcher, W.R. 1994. Riverine fisheries and habitat exploitation of the Indus Valley tradition: an example from Harappa, Pakistan, in A. Parpola & P. Koskikallio (ed.), *South Asian Archaeology 1993*: 71-80. Helsinki: Suomalainen Tiedeakatemia.

Bernhard, W. 1967. Human skeletal remains from the cemetery of Timargarha, *Ancient Pakistan* III: 291-407.

Bettinger, R. 1991. *Hunter-gatherers: archaeological and evolutionary theory*. New York: Plenum.

Binford, L.R. 1983. *In Pursuit of the Past. Decoding the Archaeological Record*. London: Thames & Hudson.

Binford, L.R. 1989. *Debating Archaeology*. San Diego, California: Academic Press, Inc.

Bond, J.M. 1994. *Change and Continuity in an Island System; the Palaeoeconomy of Sanday, Orkney. An integrated study of the faunal and botanical remains from multiperiod (prehistoric to Viking-Age) sites in Orkney*. Unpublished Ph.D dissertation, University of Bradford.

Bond, J.M. & O'Connor, T.P. 1999. *Bones from Medieval Deposits at 16-22 Coppergate and Other Sites in York*. York: Council for British Archaeology.

Bowman, S. 1990. *Radiocarbon Dating*. London: British Museum Publications Ltd.

Branigan, K. & Dearne, M.J. 1992. *Romano-British Cavemen. Cave Use in Roman Britain*. Oxford: Oxbow.

Brochier, J.E., Villa, P. & Giocomarra, M. 1992. Shepherds and Sediments: Geo-ethnoarchaeology of Pastoral Sites, *Journal of Anthropological Archaeology* 11:47-102

Callieri, P. 1992. Bir-kot-ghwandai: An Early Historic Town in Swat (Pakistan), in C. Jarrige (ed.), *South Asian Archaeology 1989*: 339-346. Madison, Wisconsin: Prehistory Press.

Caloi, L. & Compagnoni, B. 1976. Bone remains from Loebanr III (Swat, Pakistan) *East and West* 26: 31-43.

Caroe, O. 1958. *The Pathans. 550 B.C. - A.D. 1957*. Oxford: Oxford University Press.

Chang, C. & Tourtellotte, P.A. 1993. Ethnoarchaeological Survey of Pastoral Transhumance Sites in the Grevena Region, Greece, *Journal of Field Archaeology* 20:249-264.

Childe, V.G. 1957. Civilizations, Cities, and Towns, *Antiquity* 31: 36-8.

Collier, J. & Collier, M. 1986. *Visual Anthropology. Photography as a Research Method*. Albuquerque: University of New Mexico Press.

Compagnoni, B. 1979. Preliminary report on the faunal remains from protohistoric settlements of Swat, in M. Taddei (ed.), *South Asian Archaeology 1977*: 697-700. Naples: Istituto Universitario Orientale.

Compagnoni, B. 1987. Faunal Remains,. in G. Stacul (ed.), *Prehistoric and Protohistoric Swat, Pakistan (c. 3000 - 1400 B.C.)*: 131-154. Rome: IsMEO.

Coningham, R.A.E. & Sutherland, T.L. 1998. Dwellings or granaries? The pit phenomenon of the Kashmir-Swat Neolithic, *Ancient Pakistan* XII: 177-187.

Coningham, R. & Young, R. 1999. The archaeological visibility of caste: an introduction, in T. Insoll (ed.), *Case Studies in World Religion*: 84-93. Oxford: BAR International Series 755.

Coningham, R.A.E. 1995. Dark Age or continuum? An archaeological analysis of the second emergence of urbanism in South Asia, in F.R. Allchin (ed.), *The Archaeology of Early Historic South Asia*: 54-72. Cambridge: Cambridge University Press.

Costantini, L. 1979. Notes on the palaeoethnobotany of protohistorical Swat, in M. Taddei (ed.), *South Asian Archaeology 1977*: 703-708. Naples: Istituto Unviersitario Orientale.

Costantini, L. 1987. Vegetal Remains, in G. Stacul (ed.), *Prehistoric and Protohistoric Swat, Pakistan (c. 3000 - 1400 B.C.)*: 155-165. Rome: IsMEO.

Cresser, M., Killhan, K., Edwards, T. 1993. *Soil chemistry and its applications*. Cambridge: Cambridge University Press.

Cribb, R. 1987. The logic of the herd: a computer simulation of archaeological herd structure, *Journal of Anthropological Archaeology* 6: 376-415.

Cribb, R. 1991. Mobile villagers: the structure and organisation of nomadic pastoral campsites in the Near East, in C.S. Gamble & W.A. Boismier (ed.), *Ethnoarchaeological Approaches to Mobile Campsites. Hunter-Gatherer and Pastoralist Case Studies*: 371-393. Ann Arbor, International Monographs in Prehistory.

Dani, A.H. (ed.) 1967. Timargarha and Gandhara Grave Culture, *Ancient Pakistan* III.

Dani, A.H. 1968-9. Introduction, *Ancient Pakistan* IV:1-32

Dani, A.H. 1992. Pastoral-agricultural tribes of Pakistan in the post-Indus period, in A.H. Dani & V.M. Masson (ed.), *History of civilizations of Central Asia*: 395-419. Paris: UNESCO Publishing.

Dani, A.H. 1986. *The Historic City of Taxila*. Paris: UNESCO.

Davis, S.J.M. 1987. *The Archaeology of Animals*. London: Batsford.

Dennell, R.W. 1972. The interpretation of plant remains: Bulgaria, in E.S. Higgs (ed.), *Papers in Economic Prehistory*: 149-159. Cambridge: Cambridge University Press.

Dennell, R. 1983. *European Economic Prehistory. A new approach*. London: Academic Press.

Dichter, D. 1967. *The North-West Frontier Province of West Pakistan. A Study in Regional Geography*. Oxford: Clarendon Press.

Durrani, F.A. 1988. Excavations in the Gomal Valley. Rehman Dheri excavations, *Ancient Pakistan* VI: 1-232.

Durrani, F.A., Qamar, M.S. & Khan, S.N. 1997. Preliminary Report on Excavations at Manek Rai Dheri off Panian Sarikot Road, Haripur Valley, *Athariyyat* 1: 213-232.

Edlin, H., Nimmo, M., Paterson, A., Morris, P., Burgis, M., Mason, J., Pitt, J., Palmer, J., Paterson, R., Hepper, F., Dransfield, J., & Melville, R. 1978. *The Illustrated Encyclopedia of Trees. Timbers and Forests of the World*. London: Salamander Books Ltd.

Ellen, R.F. (ed.). 1984. *Ethnographic research. A guide to general conduct*. London: Academic Press.

Erdosy, G. 1995. Language, material culture and ethnicity: theoretical perspectives, in G. Erdosy (ed.), *The Indo-Aryans of Ancient South Asia. Language, Material Culture and Ethnicity*: 1-31. Berlin, Walter de Gruyter.

Faccenna, D. 1964. *A guide to the excavations in Swat (Pakistan)(1956-1962)*. Italian Mission.

Fuller, Dorian Q., in press. Fifty years of archaeobotanical studies in India: Laying a solid foundation, in S. Settar and R. Korisettar (ed.), *Indian Archaeology in Retrospect, Volume III. Archaeology and Interactive Disciplines*. New Delhi: Oxford & IBH Publishing Co.

Fussman, G. 1993. Taxila: The Central Asian Connection, in H. Spodek & D.M. Srinivasan (ed.), *Urban Form and Meaning in South Asia: The Shaping of Cities from Prehistoric to Precolonial Times*: 83-100. Washington: National Gallery of Art.

Geddes, D.S. 1983. Neolithic transhumance in the Mediterranean Pyrenees, *World Archaeology* 15, 1: 51-66.

Gilbert, A.S. 1983. On the origins of specialized nomadic pastoralism in western Iran, *World Archaeology* 15, 1: 105-119.

Gilbertson, D.D., & Hunt, C.O. 1996. Romano-Libyan Agriculture: Walls and Floodwater Farming, in G. Barker (ed.), *Farming the Desert. The UNESCO Libyan Valleys Archaeological Survey*: 191-225. Tripoli & London: UNESCO.

Gordon, D.H. 1960. *The Pre-Historic Background of Indian Culture*. New Delhi: Munshiram Manoharlal Publishers Pvt. Ltd.

Government of Pakistan. 1994. *1990 Census of Agriculture. Province Report. N.W.F.P. (Including Frontier Regions and Agencies)*. Lahore, Pakistan: Agricultural Census Organization, Government of Pakistan.

Grantham, B. 1995. Dinner in Buqata: the symbolic nature of food animals and meal sharing in a Druze village, in K. Ryan & P.J. Crabtree (ed.), *The symbolic role of animals in Archaeology*: 73-78. Pennsylvania: MASCA.

Halstead, P. 1987. Traditional and Ancient Rural Economy in Mediterranean Europe: Plus Ca Change? *Journal of Hellenistic Studies* CVII: 77-87.

Halstead, P. 1990. Present to Past in the Pindhos: diversification and specialisation in mountain economies, *Rivista di Studi Liguri* LVI 1-4: 61-80.

Halstead, P. 1992. From reciprocity to redistribution: modelling the exchange of livestock in Neolithic Greece, *Anthropozoologica* 16:19-30.

Hammersley, M. 1992. *What's wrong with ethnography? Methodological explorations*. London & New York: Routledge.

Haserodt, K. 1996. The geographical features and problems of Chitral: a short introduction, in E. Bashir & Israr-ur-Din (eds.), *Proceedings of the second international Hindu Kush cultural conference*: 3-18. Karachi: Oxford University Press.

Helms, S.W. 1982. Excavations at 'The city and the famous fortress of Kandahar, the foremost place in all of Asia' *Afghan Studies* 3-4: 1-24.

Hesse, B. & Wapnish, P. 1985. *Animal Bone Archaeology*. Washington: Taraxacum.

Higham, C. 1975. *Non Nok Tha. The Faunal Remains From The 1966 and 1968 Excavations At Non Nok Tha, Northeastern Thailand*. Otago: Otago University Studies in Prehistoric Anthropology.

Hillman, G. 1984. Interpretation of archaeological plant remains: The application of ethnographic models from Turkey, in W. Van Zeist & W.A. Casparie (ed.), *Plants and Ancient Man. Studies in Palaeoethnobotany*: 1-41. Rotterdam: A.A. Balkema.

Hillman, G. 1981. Reconstructing crop husbandry practices from charred remains of crops, in R. Mercer (ed.), *Farming Practices in British Prehistory*: 123-162. Edinburgh: Edinburgh University Press.

Hillson, S. 1992. *Mammal Bones and Teeth*. London: University College London.

Hillson, S. 1986. *Teeth*. Cambridge: Cambridge University Press.

Hodder, I. 1999. *The Archaeological Process*. Oxford: Blackwell Publishers.

Hodder, I. 1982. *The Past Present*. London: Batsford.

Hubbard, R.N.L.B., & Clapham, A. 1992. Quantifying macroscopic plant remains. *Review of Palaeobotany and Palynology*. 73, 117-132.

Imperial Gazetteer of India. 1904. *North-West Frontier Province*. Lahore: Sang-E-Meel Publications.

Insoll, T. 1999. *The Archaeology of Islam*. Oxford: Blackwell Publishers.

Jarrige, J-F. 1997. From Naushero to Pirak: Continuity and Change in the Kachi/Bolan Region from the 3rd to the 2nd Millennium B.C., in R. Allchin & B. Allchin (ed.), *South Asian Archaeology, 1995*: 11-32. New Delhi: Mohan Primlani, Oxford & IBH Publishing Co. Pvt. Ltd.

Jawad, A. 1998. Faunal remains from Kalako-deray, Swat (Mid-2nd Millennium B.C.), *East and West* 48: 265-290.

Johnson, M. *Archaeological Theory. An Introduction*. Oxford: Blackwell.

Jones, G.E.M. 1984. Interpretation of archaeological plant remains: ethnographic models from Greece, in W. Van Zeist & W.A. Casparie (ed.), *Plants and Ancient Man. Studies in Palaeoethnobotany*: 43-61. Rotterdam: A.A. Balkema.

Kenoyer, J.M. 1998. *Ancient Cities of the Indus Valley Civilization*. Karachi, Oxford: Oxford University Press.

Kenoyer, J.M. 1991. Shell working in the Indus Civilisation, in M. Jansen, M. Mulloy & G. Urban (ed.), *Forgotten Cities on the Indus*: 216-219. Mainz, Germany: Verlag Philipp von Zabern.

Khan, F., Knox, J.R. & Thomas, K.D. 1991. *Explorations and Excavations in Bannu District, North-West Frontier Province, Pakistan, 1985-1988*. London: British Museum.

Khan, S.N. 1993. Preliminary excavation report of a megalithic burial site near Adina, District Swabi, *Ancient Pakistan* VIII: 161-80.

Khoury, P.S. & Kostiner, J. 1991. Introduction: tribes and the complexities of state formation in the Middle East, in P.S. Khoury & J. Kostiner (eds.), *Tribes and State Formation in the Middle East*: 1-22. London: I.B. Tauris & Co. Ltd.

Klein, R.G. & Cruz-Uribe, K. 1984. *The Analysis of Animal Bones from Archaeological Sites*. Chicago: University of Chicago Press.

Kramer, C. 1979. Introduction, in C. Kramer (ed.). *Ethnoarchaeology: implications of ethnography for archaeology*: 1-20. New York: Columbia University Press.

Kureshy, K.V. 1977. *A geography of Pakistan*. Karachi, London: Oxford University Press.

Kuzmina, E.E. 1997. The cultural connections between the shepherds of the steppes and South Central Asia, Afghanistan and India in the Bronze Age, in R. Allchin & B. Allchin (ed.), *South Asian Archaeology, 1995*: 279-289. New Delhi: Mohan Primlani, Oxford & IBH Publishing Co. Pvt. Ltd.

Lees, S.H. & Bates, D.G. 1974. The Origins of Specialized Nomadic Pastoralism: a Systemic Model, *American Antiquity* 39, 2:187-193.

Leshnik, L.S. 1974. Land use and ecological factors in prehistoric North-West India, in J.E. Van Lohuizen-Deleeuw & J.M.M. Ubaghs (ed.), *South Asian Archaeology 1973*: 67-84. Leiden: E.J. Brill.

Lindholm, C. 1985. Models of segmentary political action: the examples of Swat and Dir, NWFP, Pakistan, in S. Pastner & L. Flam (ed.), *Anthropology in Pakistan: Recent Socio-Cultural and Archaeological Perspectives*: 21-39. Karachi: Indus Publications.

Lyman, R.L. 1994. *Vertebrate Taphonomy*. Cambridge: Cambridge University Press.

Mabberley, D.J. 1987. *The Plant Book. A Portable Dictionary of the Higher Plants*. Cambridge: Cambridge University Press.

Maclean, R. & Insoll, T. 1999. The social context of food technology in Iron Age Gao, Mali, *World Archaeology* 31(1):78-92.

Marshall, J. 1951. *Taxila*. Delhi: Motilal Banarsidass.

Marshall, J. 1904. Excavations at Charsadda. *Annual Report of the Archaeological Survey of India*: 141-184.

McMahon, A.H. & Ramsay, D.G. 1901. *Report on the tribes of Dir, Swat and Bajour*. Calcutta.

McNairn, B. 1980, *The Method and Theory of V. Gordon Childe*. Edinburgh: Edinburgh University Press.

Meadow, R.H. 1984. Notes on the faunal remains from Mehrgarh, with a focus on cattle (*Bos*), in B. Allchin (ed.), *South Asian Archaeology 1979*: 34-40. Cambridge: Cambridge University Press.

Meadow, R.H. 1989. Continuity and change in the agriculture of the Greater Indus Valley: palaeoethnobotanical and zooarchaeological evidence, in J.M. Kenoyer (ed.), *Old Problems and New Perspectives in the Archaeology of South Asia*: 61-74. Wisconsin: University of Wisconsin.

Meadow, R.H. 1991. Faunal remains and urbanism at Harappa, in R.H. Meadow (ed.), *Harappa Excavations 1986-1990. A Multidisciplinary Approach to third Millennium Urbanism*: 89-106. Wisconsin: Prehistory Press.

Mian, N.I. 1955. *A report on the cost of production and marketing of sugar-cane and gur (Mardan and Peshwar Districts)*. Peshawar: Board of Economic Enquiry, University of Peshawar.

Miller, H.M-L. 1991. Urban Palaeoethnobotany at Harappa, in R.H. Meadow (ed.), *Harappa Excavations 1986-1990. A Multidisciplinary Approach to Third Millennium Urbanism*: 121-126. Wisconsin: Prehistory Press.

Morrison, K.D. 1993. Supplying the City: The Role of Reservoirs in an Indian Urban Landscape, *Asian Perspectives* 32,2: 133-151.

Morrison, K.D. 1995. *Fields of Victory. Vijayanagara and the Course of Intensification*. Berkeley, California: Archaeological Resource Facility, University of California at Berkeley.

Mughal, M.R. 1994. The Harappan Nomads of Cholistan, in B. Allchin (ed.), *Living Traditions. Studies in the Ethnoarchaeology of South Asia*: 53-68. Oxford & New Delhi: Oxbow Books & Oxford & IBH Publishing Co. PVT. Ltd.

O'Connor, T. 2000. *The Archaeology of Animal Bones*. Stroud, Gloucestershire: Sutton Publishing Ltd.

Palmer, C. 1998. An exploration of the effects of crop rotation regime on modern weed floras, *Environmental Archaeology* 2: 35-48.

Parkes, P. 1987. Livestock symbolism and pastoral ideology among the Kafirs of the Hindu Kush, *Man* 22: 637-600.

Patel, A. 1997. The pastoral economy of Dholavira: a first look at animals and urban life in third millennium Kutch, in R. Allchin & B. Allchin (ed.), *South Asian Archaeology, 1995*: 101-113. New Delhi: Mohan Primlani, Oxford & IBH Publishing Co. Pvt. Ltd.

Payne, S. 1972. On the interpretation of bone samples from archaeological sites, in E.S. Higgs (ed.), *Papers in Economic Prehistory. Studies by members and Associates of the British Academy Major Research Project in the Early History of Agriculture*: 65-81. Cambridge: Cambridge University Press.

Payne, S. 1973. Kill-off patterns in sheep and goats: the mandibles from Asvan Kale, *Anatolian Studies* 23, 281-303.

Pearsall, D.M. 1989. *Paleoethnobotany: a handbook of procedures*. San Diego, California: Academic Press.

Piggott, S. 1959. *Approach to Archaeology*. Harmondsworth, England: Penguin Books Ltd.

Popper, V.S. 1988. Selecting quantitative measurements in paleoethnobotany, in C.A. Hastorf & V.S. Popper (ed.), *Current Paleoethnobotany. Analytical Methods and Cultural Interpretations of Archaeological Plant Remains*: 53-71. Chicago: University of Chicago Press.

Possehl G.L. 1979. Pastoral nomadism in the Indian civilization - an hypothesis, in M. Taddei (ed.), *South Asian Archaeology 1977*: 537-551. Naples, Instituto Universitario Orientale.

Possehl, G.L. 1994. *Radiometric dates for South Asian archaeology*. Philadelphia: University of Pennsylvania Museum.

Punjab Government. 1897-8. *Gazetteer of the Peshawar District. 1897-8*. Lahore: Sang-E-Meel Publications.

Rahman, A. 1968-9. Excavation at Damkot, *Ancient Pakistan* IV: 103-250.

Rahman, I.U. 1969. The ethnological wealth in Swat, *Pakistan Archaeology* 6: 285-300.

Rajan, K. 1994. *Archaeology of Tamilnadu (Kongu Country)*. Delhi: Book India Publishing Co.

Reddy, S.N. 1991. Complementary approaches to Late Harappan subsistence: an example from Oriyo Timbo, in R. Meadow (ed.), *Harappa Excavations 1986-1990. A Multidisciplinary Approach to the Third Millenium Urbanism*: 127-135. Madison: Prehistory Press.

Reddy, S.N. 1997. If the threshing floor could talk: integration of agriculture and pastoralism during the late Harappan in Gujarat, India, *Journal of Anthropological Archaeology* 16: 162-187.

Reid, A. 1996. Cattle herds and the redistribution of cattle resources, *World Archaeology* 28,1: 43-57.

Reid, A. & Young, R. 2000. Pottery Abrasion and the Preparation of African Grains, *Antiquity* 74:101-11.

Reitz, E.J. & Wing, E.S. 1999. *Zooarchaeology*. Cambridge: Cambridge University Press.

Renfrew, J. 1973. *Palaeoethnobotany*. London: Methuen & Co.

Roberts, T. J. 1997. *The Mammals of Pakistan*. Karachi: Oxford University Press.

Robinson, F. (ed.) 1989. *The Cambridge Encyclopedia of India, Pakistan, Bangladesh, Sri Lanka, Nepal, Bhutan and the Maldives*. Cambridge: Cambridge University Press.

Saraswat, K.S. 1986. Ancient crop economy of Harappans from Rohira, Punjab (c. 2,000 - 1,700 B.C.), *The Palaeobotanist* 35(1): 32-38.

Schiffer, M. (ed.) 1981. *Advances in Archaeological method and theory*. New York, Academic Press.

Schoch, W.H., Pawlik, B., & Schweingruber, F.H. 1988. *Botanical Macro-remains*. Berne: Paul Haupf Publishers.

Schopen, G. 1997. Archaeology and Protestant presumption in the study of Indian Buddhism, in G. Schopen (ed.), *Bones, Stones and Buddhist Monks: Collected Papers on the Archaeology, Epigraphy and Texts of Monastic Buddhism in India*: 1-22. Honolulu: University of Hawai'i Press.

Shaffer, J.G. 1993. Reurbanization: The Eastern Punjab and Beyond, in H. Spodek & D.M. Srinivasan (ed.), *Urban Form and Meaning in South Asia: The Shaping of Cities from Prehistoric to Precolonial Times*: 53-67. Washington: National Gallery of Art.

Shaffer J.G. & Lichtenstein, D. 1995. The concepts of "cultural tradition" and "palaeoethnicity" in South Asian archaeology. In G Erdosy (ed.) *The Indo-Aryans of Ancient South Asia. Language, Material culture and Ethnicity*: 126-154. Berlin, Walter de Gruyter.

Sharif, M. 1969. Excavations at Bhir Mound Taxila, *Pakistan Archaeology* 6: 7-99.

Shaukat, S. & Begum, R. 1992. *Rural Household Situations in Two Irrigated Villages of Peshawar Division*. Peshawar, Pakistan: Institute of Development Studies, NWFP Agricultural University, Peshawar. Publication No. 224.

Silver, I.A. 1969. The ageing of domestic animals, in D. Brothwell & E. Higgs (ed.), *Science in Archaeology. A Survey of Progress and Research*: 283-302. London: Thames & Hudson.

Spate, O.H.K. & Learmonth, T.A. 1954. *India and Pakistan. A General and Regional Geography*. London: Methuen & Co Ltd.

Stacul, G. 1966. Preliminary report on the Pre-Buddhist Necropolises in Swat (W. Pakistan) *East and West* 16: 37-79

Stacul, G. 1967a. Excavations in a rock shelter near Ghaligai (Swat, W. Pakistan). Preliminary report, *East and West* 17: 185-219.

Stacul, G. 1967b. Discovery of four pre-Buddhist cemeteries near Pach in Buner (Swat, W. Pakistan), *East and West* 17: 220-232.

Stacul, G. 1969. Excavation near Ghaligai (1968) and chronological sequence of protohistorical cultures in the Swat Valley, *East and West* 19: 44-91.

Stacul, G. 1976. Excavation at Loebanr III (Swat, Pakistan), *East and West* 26: 13-30.

Stacul, G. 1977. Dwelling and Storage-Pits at Loebanr III (Swat, Pakistan) 1976 Excavation Report, *East and West* 27: 227-253.

Stacul, G. 1978. Excavation at Bir-kot-ghundai (Swat, Pakistan), *East and West* 28: 137-150.

Stacul, G. 1979. The black-burnished ware period in the Swat Valley (c. 1700-1500 BC), in M. Taddei (ed.), *South Asian Archaeology 1977*: 661-673. Vol II. Naples: Istituto Universitario Orientale.

Stacul, G. 1980a. Bir-kot-ghundai (Swat, Pakistan): 1978 excavation report, *East and West,* 30: 55-65

Stacul, G. 1980b. Loebanr III (Swat, Pakistan): 1979 excavation report, *East and West* 30: 67-76.

Stacul, G. 1984. Cultural change in the Swat Valley and beyond, c. 3000-1400 BC, in B. Allchin (ed.), *South Asian Archaeology 1981*: 205-212. Cambridge: Cambridge University Press.

Stacul, G. 1987. *Prehistoric and Protohistoric Swat, Pakistan (c. 3000 - 1400 B.C.)*. Rome: IsMEO.

Stacul, G. 1993. Kalako-deray, Swat: 1989-1991 Excavation Report, *East and West* 43: 69-94.

Stacul, G. 1994a. Neolithic Inner Asian traditions in northern Indo-Pakistani valleys, in A. Parpola & P. Koskikallio (ed.), *South Asian Archaeology, 1993*: 708-714. Helsinki: Suomalainen Tiedeakatemia.

Stacul, G. 1994b. Querns from early Swat, in J.M. Kenoyer (ed.), *From Sumer to Meluhha: Contributions to the Archaeology of South and West Asia in Memory of George F. Dales, Jr*: 235-239. Madison, Wisconsin: Wisconsin Archaeological Reports 3.

Stacul, G. 1995. Kalako-deray, Swat: 1992-1993 Excavation Report, *East and West* 45: 109-126.

Stacul, G. 1996. Pit Structures from Early Swat, *East and West* 46: 435-439.

Stacul, G. 1997a. Early Iron Age in Swat: Development or Intrusion? in R. Allchin & B. Allchin (ed.), *South Asian Archaeology, 1995*: 341-348. New Delhi: Mohan Primlani, Oxford & IBH Publishing Co. Pvt. Ltd.

Stacul, G. 1997b. Kalako-deray, Swat: 1994 and 1996 Excavation Report, *East and West* 47: 363-378.

Stacul, G. & Tusa, S. 1977. Report on the Excavations at Aligrama (Swat, Pakistan): 1974, *East & West* 27: 151-205.

Stacul, G. & Tusa, S. 1975. Report on the excavations at Aligrama (Swat, Pakistan) 1966, 1972, *East and West* 25: 291-321.

Stein, A. 1929. *On Alexander's Track to the Indus. Personal Narrative of Explorations on the North-West Frontier of India*. London: MacMillan & Co.

Stein, M.A. 1921. *Serindia*. Oxford University Press, Oxford.

Tengberg, M. 1999. Crop husbandry at Miri Qalat, Makran, SW Pakistan (4000-2000 B.C.), *Vegetation History and Archaeobotany.* 8, 1-2: 3-12.

Thomas, K.D. 1983. Agricultural and subsistence systems of the third millennium B.C. in north-west Pakistan: a speculative outline, in M. Jones (ed.), *Integrating the Subsistence Economy*: 279-314. Oxford: BAR International 181.

Thomas, K.D. 1986a. Palaeobiological investigations, in F.R. Allchin, B. Allchin, F.A. Durrani & M.F. Khan (eds.), *Lewan and the Bannu Basin. Excavation and survey of sites and environments in North West Pakistan*: 121-136. Oxford: BAR International 310.

Thomas, K.D. 1986b. Environment and subsistence in the Bannu Basin, in F.R. Allchin, B. Allchin, F.A. Durrani & M.F. Khan (eds.), *Lewan and the Bannu Basin. Excavation and survey of sites and environments in North West Pakistan*: 13-33. Oxford: BAR International 310.

Thomas, K. 1999. Getting a life: stability and change in social and subsistence systems on the North-West Frontier, Pakistan, in later prehistory, in C. Gosden & J. Hather (ed.), *The Prehistory of Food: Appetites for Change*: 306-321. London: Routledge.

Thomas, K.D. & Knox, J.R. 1994. Routes of passage: later prehistoric settlement and exploitation of a frontier region in northwestern Pakistan, *Bulletin of the Institute of Archaeology, London* 31: 89-104.

Thomas, P.K., Joglekar, P.P., Matsushima, Y., Pawankar, S.J., & Deshpnade, A. 1997. Subsistence based on animals in the Harappan Culture of Gujarat, India, *Anthropozoologica* 25-26: 767-776.

Thompson, G.B. 1996. *The Excavation of Khok Phanom Di. A Prehistoric Site in Central Thailand. Volume IV: Subsistence and Environment: the Botanical Evidence. The Biological Remains (Part II)*. London: The Society of Antiquaries of London.

Tusa, S. 1979. The Swat Valley in the 2nd and 1st Millennia BC: a question of marginality, in M. Taddei (ed.), *South Asian Archaeology 1977*: 675-695. Naples: Istituto Universitario Orientale.

Veen, van der, M. 1992. *Crop Husbandry Regimes. An Archaeobotanical Study of Farming in Northern England 1000 BC - AD 500.* Sheffield Archaeological Monographs 3. Sheffield: J.R. Collis Publications.

Vishnu-Mittre, Sharma, A. & Chanchala 1986. Ancient Plant Economy at Daimabad, in S.A. Sali (ed.), *Daimabad 1976-79*: 588-627. New Delhi: Archaeological Survey of India.

Vogelsang, W. 1988. A period of acculturation in Ancient Gandhara. *South Asian Studies.* 4: 103-13.

Weber, S.A. 1999. Seeds of urbanism: palaeoethnobotany and the Indus Civilization. *Antiquity* 73: 813-26.

Weber, S.A. 1992. Food stress in South Asia: an explanation for culture change, in G. Possehl (ed.), *South Asian Archaeology Studies*: 253-259. New Delhi: Oxford & IBH Publishing Co. PVT. Ltd.

Weber, S.A. 1991. *Plants and Harappan Subsistence. An Example of Stability and Change from Rojdi.* New Delhi: Oxford & IBH Publishing Co. PVT. Ltd.

Weiss, A.M. 1995. The Society and its environment, in P.R. Blood (ed.), *Pakistan. A Country Study*: 75-146. Washington: Federal Research Division, Library of Congress.

Wheeler, M. 1976. *My Archaeological Mission to India and Pakistan.* London: Thames & Hudson.

Wheeler, R.E.M. 1962. *Charsadda, a metropolis of the North-West Frontier.* London.

Wilber, D.N. 1964. *Pakistan. It's People, it's Society, it's Culture.* New Haven, Connecticut: Hraf Press.

Young, R.L. 1997. *The Animal Bones from Hund, Pakistan.* Unpublished report, archived at the University of Bradford.

Young, R.L. 1998. A cattle skeleton from Gor Khuttree. *Ancient Pakistan.* XII: 139-145.

Young, R.L. & Coningham, R.A.E. 1998. Subsistence patterns in NWFP: an archaeological study. The faunal remains from Charsadda VI, Hund and Rehman Dheri. *Ancient Pakistan* XII: 93-126.

Young, R. & Nicholson, R. 1998-9. Soil pH, in J. Bond & S.J. Dockrill (ed.), *Old Scatness Broch and Jarlshof Environs Project: Interim Report No. 4: 1998-9.* (Data Structure Report) Shetland Amenity Trust & University of Bradford. ISSN 1361-3316.

Young, R., Coningham, R., Batt, C. & Ali, I. 2000. A comparison of Kalasha and Kho subsistence patterns in Chitral, NWFP, Pakistan, *South Asian Studies*: 133-142.

Zohary, D. & Hopf, M. 1988. *Domestication of plants in the Old World.* Oxford: Clarendon Press.

APPENDICES

Appendix 1 The Animal Bone Material from the Bala Hisar of Charsadda

Bala Hisar of Charsadda: Animal bones from in situ contexts
Trenches VIII & VI

Trench	Phase	Context	Species	Element	Fus/Unf	Age Est	L/R	Wght (g)	
VIII	A	1077	Bos sp.	trochanter frag	f	adult	r	32,2	
			Bos sp.	metapodial, distal artic	f	adult	l	13,1	cut
			Ovis-capra	3rd phalange frag	f	adult	r	3,5	
			Bos sp.	caudal vertebrae	f	adult		10	
VIII	A	1071	Cervid sp.	distal metacarpal frag	f	adult	r	8,3	
VIII	B	1076	Bubalus sp	scapula, glenoid cavity frag	f	adult	r	45,5	
VI	B	57	Bos sp.	M2, lower		wear = N	l	13,8	
			Bos sp.	metatarsal, proximal epiphysis	unf	juvenile	l	5,8	
			Bos sp.	femur, distal frag	f	adult	r	27,1	
			Bos sp.	2nd phalange	f	adult	l	12,9	
			Bos sp.	M3, lower		wear = F	l	27	
			Bos sp.	mandible frag			r	62	
			Bos sp.	P4, lower		wear = H	r	5,9	
			Bos sp.	M3, lower		wear = E	r	38,2	
			Bos sp.	M3, lower		wear = G	r	18,7	
			Bos sp.	M2, lower		wear = G	r	21,5	
			Bos sp.	M1, lower		wear = H	r	6,3	
			Ovis-capra	metatarsal, shaft and artic frag	f	adult	r	6,2	
			Ovis-capra	metapodial, distal artic	f	adult	l	7,8	cut
			Ovis-capra	tibia shaft frag and distal artic	f	adult	r	8,2	
			Ovis-capra	calcaneous	f	adult	r	5,4	
			Ovis-capra	M1, lower		wear = C		4,1	
			Ovis-capra	humerus	f	adult	l	23,6	
			Ovis-capra	humerus, distal frag	f	adult	r	21,3	
			Bos sp.	vertebra, neural spine	f	adult		9	
			Bubalus sp?	metatarsal, proximal frag	f	adult	r	25,4	
			Bos sp.	mandible frag			r	31,6	
			Bos sp.	mandible frag			l?	13,2	
			Ovis-capra	ulna, proximla frag	f	adult	r	4,3	
VI	B	60	Bos sp.	hunerus, distal frag	f	adult	r	100	
			Sus sp.	M2, lower		wear = E	l	3	
			Bubalus sp	2nd phalange	f	adult	r	17	burnt
			Ovis-capra	PM1 and mandible frag		wear = C	l	2	
			Sus sp.	canine frag		slight wear, juvenile?	r	1	
VI	B	71	Ovis-capra	2nd phalange	unf	juvenile	r	0,9	
			Ovis-capra	distal metacarpal artic frag	f	adult	?	2,4	
			Bos sp.	rib frag with proximal artic		adult		21,4	
			Bos sp.	tibia shaft frag and distal artic	unf	juvenile	l	62,8	
			Bos sp.	1st phalange	f	adult	r	20	
			Bos sp.	lumbar vertebra	f	adult		7,4	
			Sus sp.	proximal calcaneous frag	f	adult	?	53,6	
			Bos sp.	metatarsal distal artic frag	f	adult	r	7,6	
			Bos sp.	scapula	f	adult	l	21	
			Ovis-capra	caudal vertebra	f	adult		7,4	burnt
			Ovis-capra	ulna frag	f	adult	?	7,1	burnt
			Cervid sp.?	metatarsal proximal and shaft	f	adult	r	18	
			Ovis-capra	metatarsal frag	f	adult	r	14,2	
			Bubalus sp	distal humerus artic	f	adult	r	101	
			Bubalus sp	calcaneous	f	adult	r	72,2	
			Bubalus sp	metacarpal proximla frag	f	adult	r	31,2	
			Bos sp.	metatarsal distal frag	f	adult	r	28,9	

Charsadda: unidentifed bone: by element and size

Trench	Phase	Context	Element	Size	Comments
VIII	A	1071	artic frag	large bovid	
			artic frag	large bovid	
			shaft frag	large bovid	
			shaft frag	large bovid	
			shaft frag	large bovid	
			shaft frag	large bovid	
			shaft frag	large bovid	
			artic frag	ovis/capra	
			artic frag	ovis/capra	
			artic frag	ovis/capra	
			tooth frag	ovis/capra	
			shaft frag	ovis/capra	
			shaft frag	ovis/capra	
			shaft frag	ovis/capra	
			shaft frag	ovis/capra	
			shaft frag	ovis/capra	
VIII	A	1077	rib	large bovid	
			rib	large bovid	
			shaft frag	large bovid	
			shaft frag	large bovid	
			shaft frag	large bovid	
			shaft frag	large bovid	
			shaft frag	large bovid	
			shaft frag	large bovid	
			shaft frag	large bovid	
			shaft frag	large bovid	
			artic frag	large bovid	
			artic frag	large bovid	
			artic frag	large bovid	
			pelvis frag	ovis/capra	
			shaft frag	ovis/capra	
			shaft frag	ovis/capra	
			shaft frag	ovis/capra	
			shaft frag	ovis/capra	
			shaft frag	ovis/capra	
			artic frag	ovis/capra	
			artic frag	ovis/capra	
			artic frag	ovis/capra	
VIII	B	1076	shaft frag	large bovid	
			shaft frag	ovis/capra	
VIII	B	1063	shaft frag	large bovid	
			shaft frag	large bovid	
			shaft frag	large bovid	
VIII	B	1056	scap frag	large bovid	
			shaft frag	large bovid	
			shaft frag	large bovid	
			shaft frag	large bovid	
			artic frag	large bovid	
			artic frag	large bovid	
			artic frag	large bovid	
			artic frag	large bovid	
			shaft frag	ovis/capra	
			shaft frag	ovis/capra	
			tooth frag	ovis/capra	
			artic frag	ovis/capra	unf
VI	B	57	scap frag	large bovid	
			shaft frag	large bovid	
			shaft frag	large bovid	
			shaft frag	large bovid	
			shaft frag	large bovid	
			shaft frag	large bovid	
			shaft frag	large bovid	

Trench	Phase	Context	Element	Size	Comments
			shaft frag	large bovid	
			shaft frag	large bovid	
			rib frag	large bovid	burnt
			rib frag	large bovid	
			rib frag	large bovid	
			rib frag	large bovid	
			shaft frag	large bovid	
			shaft frag	large bovid	
			shaft frag	large bovid	
			shaft frag	large bovid	
			shaft frag	large bovid	
			shaft frag	large bovid	
			shaft frag	large bovid	
			shaft frag	large bovid	
			shaft frag	large bovid	
			shaft frag	large bovid	
			shaft frag	large bovid	
			shaft frag	large bovid	
			shaft frag	large bovid	
			shaft frag	large bovid	
			shaft frag	large bovid	
			artic frag	large bovid	
			artic frag	large bovid	
			artic frag	large bovid	
			artic frag	large bovid	
			tooth frag	large bovid	
			tooth frag	large bovid	
			tooth frag	large bovid	
			tooth frag	large bovid	
			rib frag	large bovid	
			rib frag	large bovid	
			shaft frag	ovis/capra	
			shaft frag	ovis/capra	
			shaft frag	ovis/capra	
			shaft frag	ovis/capra	
			shaft frag	ovis/capra	
			shaft frag	ovis/capra	
			shaft frag	ovis/capra	
			shaft frag	ovis/capra	
			shaft frag	ovis/capra	
			shaft frag	ovis/capra	
			shaft frag	ovis/capra	
			shaft frag	ovis/capra	
			artic frag	ovis/capra	
			artic frag	ovis/capra	
			artic frag	ovis/capra	
			tooth frag	ovis/capra	
			rib frag	ovis/capra	
VI	B	60	shaft frag	large bovid	
			shaft frag	large bovid	
			shaft frag	large bovid	
			shaft frag	large bovid	
			shaft frag	large bovid	
			tooth frag	large bovid	
			tooth frag	large bovid	
			tooth frag	large bovid	
			rib frag	large bovid	
			rib frag	large bovid	
			rib frag	large bovid	
			vert frag	large bovid	
			mandible frags	large bovid	
			mandible frags	large bovid	
			artic frag	large bovid	
			artic frag	large bovid	

Trench	Phase	Context	Element	Size	Comments
			shaft frag	large bovid	
			shaft frag	large bovid	
			shaft frag	large bovid	
			shaft frag	large bovid	
			shaft frag	large bovid	
			shaft frag	sus	
			shaft frag	ovis/capra	
			shaft frag	ovis/capra	
			shaft frag	ovis/capra	
			shaft frag	ovis/capra	
			tooth frag	ovis/capra	
			rib frag	ovis/capra	
			rib frag	ovis/capra	
			vert frag	ovis/capra	
			mandible frags	ovis/capra	
			artic frag	ovis/capra	
			artic frag	ovis/capra	
			shaft frag	ovis/capra	
			shaft frag	ovis/capra	
			shaft frag	ovis/capra	
			shaft frag	ovis/capra	
			shaft frag	ovis/capra	
VI	B	71	rib frag	large bovid	
			rib frag	large bovid	burnt
			rib frag	large bovid	burnt
			rib frag	large bovid	burnt
			rib frag	large bovid	burnt
			artic frag	large bovid	
			artic frag	large bovid	
			artic frag	large bovid	
			artic frag	large bovid	
			artic frag	large bovid	
			artic frag	large bovid	
			artic frag	large bovid	
			shaft frag	large bovid	burnt
			shaft frag	large bovid	burnt
			shaft frag	large bovid	
			shaft frag	large bovid	
			shaft frag	large bovid	
			shaft frag	large bovid	
			shaft frag	large bovid	
			shaft frag	large bovid	
			shaft frag	large bovid	
			shaft frag	large bovid	
			shaft frag	large bovid	
			shaft frag	large bovid	
			shaft frag	large bovid	
			shaft frag	large bovid	burnt
			shaft frag	large bovid	burnt
			rib frag	large bovid	burnt
			artic frag	large bovid	
			artic frag	large bovid	
			pelvis frag	large bovid	
			shaft frag	large bovid	
			shaft frag	large bovid	
			vert frag	large bovid	burnt
			skull frag	large bovid?	
			artic frag	sus	
			shaft frag	sus	
			rib frag	ovis/capra	burnt
			rib frag	ovis/capra	burnt
			artic frag	ovis/capra	
			artic frag	ovis/capra	

Trench	Phase	Context	Element	Size	Comments
			artic frag	ovis/capra	
			shaft frag	ovis/capra	
			shaft frag	ovis/capra	
			shaft frag	ovis/capra	
			shaft frag	ovis/capra	
			shaft frag	ovis/capra	
			shaft frag	ovis/capra	
			shaft frag	ovis/capra	
			shaft frag	ovis/capra	burnt
			shaft frag	ovis/capra	burnt
			shaft frag	ovis/capra	
			scap frag	ovis/capra	
			artic frag	ovis/capra	
			artic frag	ovis/capra	
			artic frag	ovis/capra	
VIII	C	1052	shaft frag	large bovid	
			artic frag	large bovid	
			artic frag	large bovid	
			artic frag	ovis/capra	
			pelvis frag	large bovid	
VI	C	72	artic frag	large bovid	
			artic frag	large bovid	
			artic frag	large bovid	
			artic frag	large bovid	
			artic frag	large bovid	
			scap frag	large bovid	
			tooth frag	large bovid	
			tooth frag	large bovid	
			tooth frag	large bovid	
			tooth frag	large bovid	
			shaft frag	large bovid	burnt & cut
			shaft frag	large bovid	burnt
			shaft frag	large bovid	burnt
			shaft frag	large bovid	
			shaft frag	large bovid	
			shaft frag	large bovid	
			shaft frag	large bovid	
			shaft frag	large bovid	
			shaft frag	large bovid	
			artic frag	ovis/capra	
			artic frag	ovis/capra	
			artic frag	ovis/capra	
			scap frag	ovis/capra	
			vert frag	ovis/capra	burnt
			tooth frag	ovis/capra	
			tooth frag	ovis/capra	
			tooth frag	ovis/capra	
			shaft frag	ovis/capra	
			shaft frag	ovis/capra	
			shaft frag	ovis/capra	

Appendix 2 The Plant Material from the Bala Hisar of Charsadda

Phase	^{14}C date (BC)	Context & Description	Identified material*	Number
A	1380-1090	1065 burnt layer	wheat (*Triticum* spp. L.)	4
		1063 burnt layer	lentil (*Lens culinaris* L.)	2
C	800-200/ 770-410	1031 occupation layer	wheat (*Triticum* spp. L.)	3
			weeds: goosefoot/fat hen (*Chenopodium* spp. L.)	2
			dock (*Rumex* spp. L.)	1
C		1032 burnt layer	lentil (*Lens culinaris* L.)	5
			rice (husk frag) (*Oryza* spp. L.)	4
			weeds: goosefoot/ fat hen (*Chenopodium* spp. L.)	2
			bedstraw (*Galium* spp. L.)	1
C		1043 fill of pit	wheat (*Triticum* spp. L.)	3
		1052 levelling / floor foundation	wheat (*Triticum* spp. L.)	2
			lentil (*Lens culinaris* L.)	2

* all material are seeds or grains, unless otherwise noted

Appendix 3 Animal Remains from the Northern Valleys

1 Kalako-deray

1.1 Summary of Identified Animal Bones from Kalako-deray Pits (PIV)

Taxa	Total No Bones	NISP	MNI
Bos sp.	61 (46.5 %)	53	5
Ovis/Capra	37 (38%)	34	9
Large Bovids	13 (10%)		
Small Bovids	5 (4%)		
Ovis/Capra/Gazelle	5 (4%)	5	3
Cervids	4 (3%)	4	4
Sus scrofa	3 (2%)	3	1
Carnivores	2 (1.5%)		
Birds	1 (0.8%)		
Total no Fragments	131		

(source: Jawad 1998, 268)

1.2 Elements: *Bos* sp.

Element	Right	Left	Centre	Not Known
Maxilla	1			1
Mandible	1	1		
Horn-core	3	2		
Atlas			3	
Scapula	1		1	
Humerus (distal)	3		3	
Radius (proximal)	1			
Ulna (proximal)	2			1
Metacarpal	1	2		1
Pelvis		2		
Femur (proximal)	2	2		
Femur (distal)	1	2		
Tibia (proximal)	2			
Calcaneus (proximal)				1
Talus		1		
Tarsal				2
Metatarsal				2
Metapodial				2
Phalange (1st)	3	2		
Phalange (2nd)	1			
Phalange (3rd)	1	1		
Number of sub-adult specimens:	4			

(source: Jawad 1998, 275)

1.3 Elements: *Ovis/Capra*

Element	Right	Left	Centre	Not Known
Mandible	7	7		
Horncore	1	1		
Scapula	1			
Humerus (proximal)	1			
Humerus (distal)	1			
Humerus (complete)	1			
Radius/Ulna (proximal)		1		
Radius/Ulna (complete)		1		
Ulna	1			
Metacarpal	2	1		
Femur (proximal)	1			
Femur (distal)	1			
Tibia (distal)	3			
Pelvis	1			
Calcaneus	1			
Number of sub-adult specimens:	7			

(source: Jawad 1998, 276)

1.4 Elements: *Sus Scrofa*

Element	Right	Left	Centre	Not Known
Mandible	1	1		
Maxilla	1			

(source: Jawad 1998, 277)

1.5 Age Estimation on the Basis of Tooth Eruption & Wear: *Sus scrofa*

	P2	P3	P4	M1	M2	M3
specimen 91-29					e	b
specimen 91-30			u/w	u/w	e	b
specimen 91-31				e	b	not erupted

Interpretation:
    91-29   adult
    91-30   adult
    91-31   young: 18-21 months

(source: Jawad 1998, 277-8)

1.6 Elements: *Ovis/Capra/Gazelle*

Element	Right	Left	Centre	Not Known
Humerus (distal)	1			
Metatarsals	1	2		

(source: Jawad 1998, 278)

1.7 Elements: *Muntiacus muntjak*

Element	Right	Left	Centre	Not Known
Horncore				1

(source: Jawad, 1998, 279-80)

1.8 Elements: *Naemorhedus goral*

Element	Right	Left	Centre	Not Known
Horncore				1

(source: Jawad, 1998, 280)

1.9 Elements: *Cervus* sp.

Element	Right	Left	Centre	Not Known
Antler				2

(source: Jawad, 1998, 281)

2 Aligrama

2.1 Summary of Identified Animal Bones from Aligrama (PIV-V)

	Period IV				Period IV-V				Period V			
Taxa	NISP	(%)	MNI	(%)	NISP	(%)	MNI	(%)	NISP	(%)	MNI	(%)
Felis sp.	1	4.3	1	12.5								
Canis sp.					1	1.1	1	4.8	9	1.3	3	3.2
Cervid sp.									1	0.1	1	1.1
Equus asinus/caballus	1	4.3	1	12.5	8	8.6	4	19.0	66	9.6	9	20.2
Sus scrofa	1	4.3	1	12.5	5	5.4	3	14.3	39	5.7	14	14.9
Bubalus bubalus									5	0.7	3	3.2
Bos indicus	16	69.6	3	37.5	61	65.6	8	38.1	410	59.4	38	40.4
Ovis/Capra	4	17.4	2	25.	18	19.4	5	23.8	160	23.2	26	27.7

(source: Compagnoni 1979, 701-2)

3 Ghaligai, Loebanr III & Bir-kot-ghundai

3.1 Summary of Identified Animal Bones from Ghaligai, Loebanr III & Bir-kot-ghundai

	Ghaligai			Loebanr III IV				Bir-kot-ghundai IV			
Taxa	I	II	III	NISP	(%)	MNI	(%)	NISP	(%)	MNI	(%)
Felis sp.				1	0.1	1	1.0				
Panthera sp.				1	0.1	1	1.0				
Muntiacus m.	1			2	0.2	1	1.0				
Axis porcinus	1			1	0.1	1	1.0				
Cervus sp.				2	0.2	2	2.0				
Noemorhedus goral				1	0.1	1	1.0				
Capra falconeri				3	0.3	1	1.0				
Lepus sp.				1	0.1	1	1.0				
Hystrix indica		1		3	0.3	2	2.0				
Canis familiaris			2	2	0.2	1	1.0	20	1.0	4	3.8
Equus asinus/caballus			1	5	0.4	3	3.1	158	7.7	11	10.4
Sus scrofa				43	3.6	12	12.4	105	5.1	13	12.3
Bos indicus	2	4	4	717	60.8	22	22.7	1464	71.5	46	43.4
Ovis/Capra		1	4	398	33.7	48	49.5	300	14.7	32	30.2

(source: Compagnoni, 1987, 134)

3.2 Elements: *Bos indicus*

Element	Ghaligai I	Ghaligai II	Ghaligai III	Loebanr III IV	Bir-kot-ghundai IV
Horncore				2	16
Horncore + skull				3	
Skull frags				84	26
Maxilla				16	13
Mandible	1		1	53	90
Upper teeth			1	38	218
Lower teeth	1	3		46	214
Joid					4
Atlas				5	7
Axis				2	14
Cervical vert				3	
Thoracic vert				23	2
Lumbar vert				11	1
Sacrum				3	3
Caudal vert			1	4	17
Scapula				20	20
Humerus (proximal)				2	1
Humerus (shaft)				8	
Humerus (distal)				20	38
Radius (proximal)				18	31
Radius (shaft)				6	2
Radius (distal)				7	11
Radius-Ulna				2	2
Ulna				12	27
Carpals			1	25	38
Metacarpal					1
Metacarpal (proximal)				13	87
Metacarpal (shaft)				3	5
Metacarpal (distal)				10	42
Proximal phalanx				43	74
Middle phalanx		1		27	89
Terminal phalanx				11	23
Pelvis				31	46
Femur (proximal)				8	19
Femur (shaft)				4	5
Femur (distal)				5	10
Patella				2	6
Tibia (proximal)				3	9
Tibia (shaft)				2	
Tibia (distal)				18	24
Malleolar					3
Calcaneum				24	21
Astragalus				35	37
Other tarsals				11	25
Metatarsal				1	2
Metatarsal (proximal)				18	57
Metatarsal (shaft)				6	8
Metatarsal (distal)				10	52
Metapodial frags				16	21
Sesamoid					3

(source: Compagnoni 1987, 143)

3.3 Elements: *Ovis* & *Capra*

Element	Ghaligai II	Ghaligai III Ovis/Capra	Loebanr III O/C	Loebanr III O	Loebanr III C	Bir-kot-ghundai O/C	Bir-kot-ghundai IV O	Bir-kot-ghundai C
Horncore				2			1	
Horncore + skull			1		1			2
Skull frags			1					
Maxilla			8			2		
Mandible			91			58		
Upper teeth		2	54			55		
Lower teeth			31			67		
Atlas			2	1		1		
Axis			3	1	1	3		
Cervical vert			8					
Thoracic vert			8			1		
Lumbar vert			8			1		
Caudal vert						1		
Scapula			19			5		
Humerus (proximal)			1					
Humerus (shaft)			15			7		
Humerus (distal)			12	1	1	8	6	4
Radius (proximal)			9			1	4	3
Radius (shaft)			15			7		1
Radius (distal)						1		
Radius-Ulna								3
Ulna			8			2		
Metacarpal (proximal)				3	5	5	3	3
Metacarpal (shaft)			5			2		
Metacarpal (distal)			1	3		2	2	
Proximal phalanx			11			6		
Pelvis		2	6	2	2	4		
Femur (proximal)			2	2	2			
Femur (shaft)			1	1	1	1		
Femur (distal)			2	1	1	4		
Tibia			1					
Tibia (proximal)						2		
Tibia (shaft)			6			1		
Tibia (distal)			13			10		
Calcaneum			3	1	1		1	
Astragalus	1		7			3		
Other tarsals			1					
Metatarsal				1				
Metatarsal (proximal)			2	1	1	3	1	
Metatarsal (shaft)			1					
Metapodial frags						3		

(source: Compagnoni, 1987, 144)

3.4 Loebanr III: sheep and goat epiphyseal fusion data: estimated ages according to Silver (1969)

	Unfused	Fused	Fused total	Age Estimate
Humerus (distal)	10	9	47.4%	10 months
Tibia (distal)	13	3	24.0%	18-24 months
Metacarpus (distal)	6	3		
Femur (proximal)	4	2	33.3%	30-36 months
Radius (distal)	2	7	63%	36 months

(source: Compagnoni, 1987, 145)

3.5 Elements: *Sus scrofa domesticus*

	Loebanr III IV	Bir-kot-ghundai IV
Skull frags	5	28
Maxilla	3	16
Mandible	17	32
Loose teeth	8	6
Atlas	2	
Scapula	5	2
Humerus (shaft)	1	
Radius (proximal)	1	
Radius (shaft)	4	
Ulna	1	
Metapodials	1	4
Pelvis	1	
Tibia (shaft)	2	5
Tibia (distal)	1	
Astragalus	1	
Phalanx (middle)	2	

(source: Compagnoni 1987, 141)

3.6 Bir-kot-ghundai: Pig mandibular tooth eruption and age estimates according to Boessneck - von den Driesch (1975)

	R	L	Age Estimate
M1 not in		1	> 6 months
M1 in, M2 not in	3	7	6-12 months
M1 in, M2 erupting	1	2	c 12 months
M2 in, M3 not in		1	12-18 months
M2 in, M3 erupting	1		18-21 months

(source: Compagnoni 1987, 141)

3.7 Elements: Equids

	Ghaligai III	Loebanr IV		Bir-kot-ghundai IV		
	Equus sp.	*Equus* sp.	*Equus caballus*	*Equus* sp.	*Equus asinus*	*Equus caballus*
Skull frags				1		
Maxilla		4		2	4	1
Mandible				2	9	2
Upper teeth	1		1	14	18	6
Lower teeth				6	21	
Incisor/Canine				15		
Atlas				1		
Axis				1		
Metacarpal					4	
Metac (prox)					3	
Metac (shaft)				1		
Metac (dist)					7	
Phalanx (prox)				2	8	
Phalanx (midd)					2	1
Phalanx (term)				5	1	
Pelvis					1	2
Femur (prox)				1		
Patella					1	
Tibia (prox)					1	
Tibia (dist)					1	
Calcaneum					2	
Astragalus				1	3	
Other tarsals					1	
Metatarsal					2	
Metatarsal (prox)					2	
Metapodial (rudimentary)				1	1	

(source: Compagnoni 1987, 139)

3.8 Elements: *Canis familiaris*

	Ghaligai III	Loebanr III IV	Bir-kot-ghundai IV
Mandible			7
Upper teeth		1	
Lower teeth		1	3
Atlas			1
Axis	1		
Cervical vert	1		
Scapula			2
Humerus (distal)			2
Ulna			1
Metacarpal V			1
Tibia (distal)			1

(source: Compagnoni 1987, 137)

Identified animal bones from Ghaligai, Loebanr III, Bir-kot-ghundai, Aligrama & Kalako-deray

	Ghaligai			Loebanr III		Bir-kot-ghundai	
period	I	II	III	IV		IV	
	NISP			NISP	MNI	NISP	MNI
WILD							
Felis sp.				1	1		
Panthera sp.				1	1		
Muntiacus muntjak	1			2	1		
Axis porcinus	1			1	1		
Cervus sp.				2	2		
Noemorhedus goral				1	1		
Capra falconeri				3	1		
Hystrix indica	1			3	2		
DOMESTIC							
Canis familiaris			2	2	1	20	4
Equus asinus/caballus			1	5	3	158	11
Sus scrofa domesticus				43	12	105	13
Bos indicus	2	4	7	717	22	1464	46
Bubalus bubalus							
Capra hircus/Ovis aries		1	4	398	48	300	32

	Aligrama						Kalako-deray	
period	IV		IV-V		V		IV	
	NISP	MNI	NISP	MNI	NISP	MNI	NISP	MNI
WILD								
Felis sp.	1	1						
Panthera sp.								
Muntiacus muntjak							1	1
Axis porcinus								
Cervus sp.					1	1	2	2
Noemorhedus goral							1	1
Capra falconeri								
Hystrix indica								
DOMESTIC								
Canis familiaris	1	1	9	3				
Equus asinus/caballus	1	1	8	4	66	9		
Sus scrofa domesticus	1	1	5	3	39	14	3	1
Bos indicus	16	3	61	8	410	38	61	5
Bubalus bubalus	5	3						
Capra hircus/Ovis aries	4	2	18	5	160	26	42	12

sources: Compagnoni 1987, 1979; Jawad 1998

Appendix 4 Plant Remains from the Northern Valleys

Palaeobotanical Evidence from Ghaligai, Bir-kot-ghundai and Loebanr III

	Ghaligai I 3000-2500BC	Ghaligai II 2400-1900BC	Ghaligai III 1900-1700BC	Loebanr III IV 1700-1400BC	Bir-kot-ghundai IV 1700-1400BC
Cereals					
Triticum aestivum					12
T. sphaerococcum		2		17	
Triticum sp.		14			
Oryza sativa				9	39
Hordeum distichum				16	
Hordeum vulgare			1	111	28
H. vulgare nudum				16	
Hordeum sp.					9
Legumes					
Lens culinaris				64	
Pisum sativum				1	
Oil seeds					
Linum usitatissimum				11	
Fruits					
Celtis australis	***	***			
Vitis vinifera				2	
Weeds					
Aegilops sp.				4	
Argemone sp.				2	
Euphorbia elioscopia		1			
Galium sp.				3	
Lithospermum arvense		1			
Secale sp.				1	

(source: Costantini 1987, 165)

Identified plant remains from Ghaligai, Loebanr III, Bir-kot-ghundai & Aligrama

	season	Ghaligai I	Ghaligai II	Ghaligai III	Loebanr III IV	Bir-kot-ghundai IV	Aligrama IV-V
CEREALS							
Triticum sp.	W		16		17	12	*
Hordeum sp.	W			1	143	37	*
Oryza sativa	S				9	39	*
LEGUMES							
Lens culinaris	W				64		*
Pisum sativum	W				1		*
Phaseolus sp.	W						*
OIL & FIBRE							
Linum usitatissimum	W				11		
FRUITS							
Celtis australis	S	*	*				*
Vitis vinifera	S				2		
WEEDS							
Aegilops sp.					4		
Argemone sp.					2		
Euphorbia elioscopia			1				
Galium sp.					3		
Lithospermum arvense			1				
Secale sp.					1		

source: Costantini 1987, 1979
* denotes species or taxon present but not quantified
W winter sowing and spring harvest (rabi crop)
S summer sowing and autumn harvest (kharif crop)

Appendix 5 The ethnographic interviews

This appendix contains summaries of each of the interviews conducted with the help of an interpreter, during the 1998 and 1999 field seasons. A set of questions to be asked in the different study regions was prepared prior to each field season. However, in practice it was not always possible to ask all of the questions included in this set, and often other information, not directly relevant to this study was either solicited by the interpreter, or volunteered by those being interviewed. Also, producing a written copy of the question set, and writing down answers, in some cases produced suspicion and uneasiness in the informants, although on other occasions the opposite effect was observed. Further, many of the interviews were conducted with herders, either moving their flocks on the road, or keeping watch over them while the animals were grazing, or in the case of the farmers, while they were working on their land. This meant that the full attention of the informants for any sustained length of time was hard to achieve. The necessity to record information directly after an interview when removed from the informant means that giving a direct transcription impossible, and so for each of the interviews conducted a summary is given, containing the information relevant to this thesis, that is, direct subsistence information, and where relevant, information relating to seasonal movement. The primary areas of focus are therefore: animals, crops, movement (covering where and when movement is made, and who moves and why). Field work was conducted with the express aim of building up a comparative picture of the subsistence and movement patterns of the inhabitants of Dir, Swat and Charsadda area, plus where possible the reasons for these choices. The interviews have been divided according to similar subsistence and/or mobility strategies, and the interpreter is noted for each interview. Where possible, the interviews have been left in words and phrases as close to the original as possible.

Interpreters:

IA is Professor Ihsan Ali, Chairman, Department of Archaeology, University of Peshawar, Peshawar, NWFP

MD is Mr Mukhtar Ali Durrani, Lecturer, Department of Archaeology, University of Peshawar, Peshawar, NWFP

QN is Mr Muhammad Qazi Naeem, Lecturer, Department of Archaeology, University of Peshawar, Peshawar, NWFP

1. Winter transhumants

1.1 Shah Muhammad and family, moving from Swabi District, Vale of Peshawar up to Kalam, Kalesh, in the northern part of Swat, interviewed near Malakand. This group comprised three families, or related households. It was not clear (either in this or in other interviews) whether 'family' groups and the numbers we were given included women and children, however, out of the three families, nine were reported still in Swabi area, while on the road there were fourteen people reported. The sheep and goats (c. 240-250) were still in Swabi, but the buffalo (eight - nine) were on the road with this group. The travelling time from Swabi to Saidhu Sharif (near Mingora) by road (on foot) is around six days, then above Saidhu trucks are used. This is because down here the roads are wider, and so suitable for buffalo. Up beyond Saidhu, the roads are narrower, and it is difficult for the buffaloes to walk, but buffaloes are 'recent' and only possible with trucks. This movement, to the same village in Swabi was done by grandfather, and family before him too. It is possible to go down to the plains for winter and leave the family home and land because at this time there is not so much land under cropping. Up in the valleys, they own between two and three acres of land in a village called Gourney which is about two kilometres from the main road through the mountains. They grow maize, potatoes, peas, and harvesting will occur one and a half months from now (June). Sowing occurred before leaving, and the land was left for the winter. The herds of sheep and goat numbers around 240-250, with more goats, fewer sheep, and there are some bulls still up in the mountains, who stay with one male relative to look after them. During the winter, snow covers all the land and there can be no work done on it. (IA)

1.2 Guful Khan and the extended family group of three sons and their own families, travelling from Mardan District, up to northern Swat, to a village called Mingal, ten kilometres south of Kalam, and about five kilometres from the main road. This whole family group travels down to Mardan, to the same village very year, then back up to Swat, to avoid the winter. They own maybe five acres of land up in Swat, and grow maize, potatoes, and wheat. One of the sons of the brothers will stay back in village, and live during winter on stored maize and wheat. He will clean the top roof free from snow to get in and out of houses. These days, the mosque is important through winter, and can provide food and fuel. No-one washes much in winter. Two cows and two bulls stay up there, and the cows are used for milk. These cows are fed on dried grass which has been collected through the summer and autumn. The bulls stay up in the village because they are used for ploughing, and unlike buffalo or goat or sheep, do not need to have good food all year round. The dried food is sufficient through the winter. They have a flock of sheep and goat, maybe 150 in number, with more goat than sheep, maybe five buffalo, and four cows (two of which stay up in the valley) and the two bulls. The snow stays for four months, November to February, and no work can be done on the land. (IA)

1.3 Gul Muhummad, a goat herder, on the main road from the Malakand Pass up to Chakdara, going to Jambil from a village four kilometres from Charsadda. Gul Muhummad had a flock of c. fifty goats, moving them from the Vale of Peshawar back up to their home. His parents already moved back up to their home, which is in a valley parallel to Dir (between Swat and Dir). Every winter, they take their buffalo, sheep and goats down to Vale, and it takes about ten days to walk from Jambil to Charsadda. They have six buffaloes, kept for milk, and both sheep and goat. They have maybe ten sheep and forty goats, and around a third of the flock on the road have kids with them, indicating that lambing has taken place down in the plains. The family has a little land up in valley, and the crops grown up there are maize, potatoes.

No-one stays up in the valleys over winter - there is nothing to stay up there for. Apart from their animals, which include two donkeys, and four cows that are up ahead with parents, they have only utensils to use on the road, cooking and eating things, and a tent. One brother in the family has adopted permanent living in Charsadda, so has a house there. (IA)

1.4 Shah Nazar, a shepherd interviewed on the common, Government owned land around the site of the Bala Hisar of Charsadda (see Plate 10, page 375). He and comes from Sherai Darra, a village in Dir (200 kilometres north of Charsadda). Shah Nazar comes to Charsadda every year, to the same place, and will return to Dir in May, after the wheat harvest in the plains, as wheat work means cash. He is here with three men, his older brothers, and their families, and these men who are also living in the same house, and do paid agricultural work, within a five kilometre radius of the Bala Hisar. They have come to the same house for many years, the landlord welcomes them. They pay rent in animal dung, used by the landlord for fuel and fertiliser. They also work in the fields and in the harvest. The flock grazing at the Bala Hisar consists of 160 sheep and goats, and maybe a quarter to a third of the animals had young. Sheep and goat travel by foot, on the roads. They always use roads, they are quicker and easier. Buffaloes and cows travel down in trucks, also the old and young people. He has twelve buffaloes (one male) and six cows (one bull). The house in Dir is tended by neighbours for the six months of winter. They have only a small plot of land in Dir, about one acre, where maize is grown, but only one crop per year. The ground is prepared after the snow melts, and is ploughed at the end of May. Sowing takes place in June, and the crop is harvested after two and a half months. They do not have a crop planted over winter. Shah Nazar's family has lived in Dir for many generations. During the trip down, which takes around ten days on foot, they stop with the flock wherever there is suitable grazing at night. While down in the Vale, they sell twenty to twenty five of their flock per season. While in Dir, only one or two according to family needs. Shah Nazar's family do not have gardens or orchards, so they don't bring down fruits, but others from their village do. They also have a temporary shed in the mountains, above the house, in the forest, and their animals are taken up there for good grazing during summer. Grazing up there they need two shepherds (two of the brothers) to guard the flock at night. Grazing is free, but there is a danger of leopards, panthers, wolf, and jackal type animal. Others, sometimes two to three groups together share the guarding, and they have special big, black ferocious dogs. Fresh food is brought up from the village. In villages in the plains with populations of up to 1,500-2,000 people, there may be a winter influx of up to fifteen to twenty groups (of seven to ten people in each) coming down each year from the valleys. These groups tend to avoid the big towns, as what they want is grazing for their flocks and somewhere they know to stay over the winter. (IA)

1.5 Sher Ali, a shepherd tending his flock by the Bala Hisar, is from Dir, in a side valley by Dir town, and has come to Charsadda for the winter, with his family, which is his father, mother and brothers, and their families. Sher Ali says his family are Gujars, and they come down to Charsadda region in October, and return in May, and the whole household comes down, no-one remains up in the valley. Now, the older people and any young people come by truck, with buffalo. They bring cows and goat by foot. Other families with more animals must go by foot. On foot the journey takes 10-20 days. Sher Ali's family have seven cows and fifty goats. (However, we did clearly see sheep and goat together, so it may be that no distinction is necessarily made between the two - but goats outnumbered sheep). If they sell any animals, it will be only those with no milk, they will keep those giving milk. They have been coming here for many years, to the same place, and give the house owner dung and dried fruits for rent. Up in their home, their village by Dir they have some small amount of land, and wheat and maize is grown up here. They scatter the seeds before they leave, and will harvest on return. Potatoes and some other vegetables are also grown. No-one stays to look after the crops, the land will have snow, and no-one goes there. (MD)

1.6 Shah Alam, herder in Karha Khwar, south east of Malam Jabba. Originally he and his family were from the village of Shering Al in Dir Kohistan, but now live in Malam Jabba permanently. He and his family are Gujars. They have no sheep, goat or buffalo, only twenty five cows and one bull. They live up here in Malam Jabba in the summer for four to five months, then in winter travel down to Charsadda. The whole family will go down, bringing animals and luggage with them. They exchange labour for wages, particularly working with the sugar crop, and they also sell milk. Milk, and milk products are the main source of cash income. They always return to the same place in Swat, their home. They do no farming, no crops are grown in either place. They do not own their land here, but pay the owner 50 rupees per cow per season rent, and also provide fertiliser to the owner for use on fields. The calves are born at no specific time, it can be either up in the valleys, or down in the plains. Older cows are sent to the butcher for money. The bull is very important, and great care is taken of him. He is moved to different pastures every day, and can sometimes be taken as much as three to five miles to find good grazing. Two people were out taking care of the cattle. Straw is needed to enhance both the quality and the quantity of milk obtained, but they cannot afford to buy it. So, alas, their cattle only eat grass and leaves which gives less milk. (QN)

1.7 Ali Haider, herder and farmer in Lak Nandai village, north of Malam Jabba. Ali Haider has a permanent home in Lak Nandai with his family. Every winter they take their animals down to Charsadda to avoid the snow and bad weather. All the family go, none remain in Lak Nandai. Ali Haider said he and his family are Gujar, and have been travelling up and down for as long as he and his family can remember. They have both sheep and goat, around eighty or ninety, and ten cows. The cows give milk, and the goats, and the milk, yoghurt and ghee is sold. They also spin and weave items for sale. They grow some crops, mainly wheat and maize, and the planting occurs before the winter, before moving down to Charsadda. They will leave their land and house, and then return in spring, and be able to harvest. The waste from wheat is given to the cows for their milk. The animals are taken out to graze around the mountains during the summer, travelling maybe four or five kilometres per day, and also in winter. They are grazed around common land or open land, and tended by the boys of the family. Their house in Lak Nandai is permanent, that is where they live. Movement to Charsadda is temporary, even though it is done every year to the same place, same village. They will help with the harvest down there, and give dung for payment to the landlord. Without buffalo they travel all on foot with their animals, and the return is made in spring, when lambs and kids might be born. Some are born at Charsadda, some in Swat, depending on the time of the move. (QN)

1.8 Taj Bar, a herder in mid Swat. Taj Bar has a house in the area of Karha Khwar, which is south east of Malam Jabba. He and his family are Gujar, and they have a flock of forty goat and sheep, and thirteen or fourteen cows, no buffalo. They live in Malam Jabba from May to September, then they move down to winter in Mardan, Peshawar or Charsadda areas. One or two family members will go ahead down to the plains to locate a place to live, and find jobs. They return to the same place in Malam Jabba every year, that is their home. They do not own the place, they pay a rent of 50 rupees per cow and 20 rupees per goat or sheep for the season of four months. The lambing season for goat and sheep is spring, and sometimes births will occur here, in Malam Jabba, and sometimes down on the plains. Spring is normally spent in Mardan. Spinning and wool is very important, it is a cash item. Milk is also sold, and other dairy products. Particularly good is the nearby tourist resort, where people will buy milk, and yoghurt for picnic lunches. Here, the animals are grazed, taken out early in the morning, they travel up to three kilometres to find good grazing. Two people, sons, go with the cows, and two with the goats. For the lambs and kids only, food is collected, it is usually grass or leaves. This family grow no crops themselves. However, they did say that those branches of their tribe who had migrated down to live in Lower Swat grew crops. Dung is collected and sold locally as a fertiliser, and the landlords also take dung as rent payment. In the plains, they provide dung as fertiliser to their landlords. Firewood is gathered, and kept inside, but dung is used when wood is not available, particularly when raining. (QN)

See Figures 5 and 6. These show the ground plan of Taj Bar's house, indicating animal and living areas. The house was constructed from dried mud, tree trunks and branches, with a mud and straw roof. There was no visible differentiation between construction technique for the animal or living areas.

1.9 Mukhtar Biland Khan, was interviewed on the road with his herd, to the south of Dir town. He and his family had spent the winter months near to Charsadda, and were travelling back up to their home, a small village in a side valley above Dir. Mukhtar Biland Khan said he was Pathan, and was a settled farmer, with a home in Dir. But every year they would travel down to Charsadda to avoid bad weather, and get grazing. The herd was c. 200 sheep and goat, and around 20 cattle. Some buffalo also were kept, but these were being taken up to the home by truck. Buffalo were recent, and recent taste for buffalo milk in the towns. Mukhtar Biland Khan and his family grew wheat and maize and vegetables at their home. The wheat would be sown on prepared ground before leaving the valley, and then it would be covered with snow all winter. They would harvest after their return home, and then grow vegetables, before planting more wheat. The men would sometimes work down near Charsadda, helping with sugar cane and wheat harvest, and they would go back to the same village. (QN)

1.10 Pervez Mirdad and his family and herds were camped in Nal village, in the Mutha Tahsil, fifteen kilometres north of Mingora, by the Swat river. The camp was on very swampy land, with both grazing and water for buffalo. There were in fact two groups camped on this area of land: one group, of Gujars was the larger, with twenty tents, and 500 metres away a smaller group with seven tents of Changa Ryan (itinerant merchants, from Peshawar, with no animals, or other land based production). This group have buffalo, two or three per family, and keep them for their milk, which they sell locally. This whole group of Gujars say they come from near to Charsadda, where they spend the winter. However, this group said that they considered themselves to belong to Swat, to a village north of Mingora, not the one camped at, but one called Gulbar, more remote than Nal. They grow no crops, they buy what they need in terms of flour from the money they get from milk, and they eat dairy products themselves. They may also gather wild green plants, and gather food for their animals. (QN)

1.11 Shah Nazar Khan, a shepherd and farmer from lower Swat. Shah Nazar Khan and his family live in the village of Gulibegh, situated half way between Malakand and Mingora, in the lower part of the Swat Valley. They keep sheep and goat, around 150, and fifteen to twenty cattle. They have no buffalo, as buffalo are only recent to Swat, and buffalo are more work than cattle or goat. Buffalo do give milk for higher money, but they need special food. Shah Nazar Khan and his family are Pathans, and their home is in Swat. Every year, in October or November, the whole family moves down to Mardan, to another village, where they go to the house they use, and they stay there until April or may, depending on the wheat harvest. This has been done for as many years as they can remember, and is to leave the winter in Swat, and to find better grass for the herds. If they were to stay in Swat, as some do, they would need to spend all summer gathering and storing food for themselves and their animals. This way they have food there, and can earn wages too. They can give animal dung to the landlord in Mardan, and they can sell other things like the wool and embroidery that has been made from their own wool. Crops are grown up in Swat. They grow wheat and rice, and maize and potatoes. Wheat is sown in September, and is left while they come down to the plains. If it snows, that will cover it, and it is ready to harvest when they return to Swat in April and May. After that, the land is prepared for rice, and irrigation is set up. The irrigation is channels running off from the river, and in summer, the rivers are full, maybe even flooding. This means that the fields can be very wet, which is needed for rice. The rice needs weeding, and will be harvested at the end of summer. Rice can be sold, and maize, but wheat and potatoes are for the family. (QN)

2. Inter-valley transhumants: winter

2.1 Banaras Ali, a farmer and animal herder, lives near Kalako-deray, in the upper Jambil Valley, Swat. Banaras Ali, who says he is Gujar, has his permanent home in the Jambil Valley, where he owns some land, maybe four acres. Every year, in the winter, he and his family take their animals down to an area just above Malakand, where they stay for five months. They do this to avoid the winter and snow, and to find grazing for their animals. They do not always go to the same place, they camp, and sometimes spend the months moving (see Plate 11, page 375). They have sheep and goat, mainly goat, around 150, and ten cows, three bulls. Banaras Ali said when ploughing was done, bulls were used. He said the land was ploughed in summer, after harvest. Wheat, maize and potato were grown on his land. Wheat was harvested in May, and then the land was prepared for the next crop. Wheat was sown in October, and left over the winter. Nothing was stored over the winter, they bought flour and food at Malakand with money from milk. (QN)

2.2 Gul Bar, a herder in Swat. Gul Bar and his immediate family live in Bishai village, Malam Jabba, although they are originally from Jan Dari village, Kal Kot Tahsil, Dir Kohistan. Gul Bar has here around sixty sheep and goat, and between fifteen to twenty cows, all of which are split into four groups and scattered, both for protection and control, and to take best advantage of grazing. This family lives in Malam Jabba in the summer, and then moves down to Takt-I-Bahi area in the winter, to the village of Fazalabad, where they have their own house. Takt-I-Bahi is just above the Malakand Pass. Thus they go to the same place every year. At Dir, they also have brothers who farm and own cows and sheep. These brothers stay up there to look after crops, but most of the animals are down here in Swat, as they need food. The brothers remaining in Dir also move down to Fazalabad for the winter, going back up to harvest the crops in May. The crops grown up in Dir are wheat and rice, with maize and potatoes too. Some rice is sold, and some wheat, but the vegetables are for their families, and also some wheat and rice. (QN)

2.3 Mian Khan, farmer and herder. Mian Khan and his family live in Upper Swat, in Tandachena Village, Chil Valley, east of Madyan. They are Gujars, and Mian Khan said his family had always lived in Swat. Their house and land is here, but every winter the family travels down to Malakand, where they stay for some months. They do not stay in one place, but move with their animals for grazing. They have sheep and goat, maybe 150, and 20 or 30 cattle. No buffalo, but perhaps three donkeys. They have crops growing over the winter, wheat and maize, that are sown before travelling down to Madyan and harvested on return. Mian Khan said these days many families stay in the village, but those that do need to collect grass for their animals, and it is easier with fewer animals to look after them up in the village through snow. Those with more animals will still travel for the winter. (QN)

2.4 Gull Faraz, a herder. Gull Faraz and his family live permanently in the village of Lal Bandi, midway between Mingora and Madyan in Swat. Gull Faraz said his family were Gujars, and may have in the long ago past come from Swat Kohistan. They have a herd of sheep, goat maybe 200 in number, and around ten cattle, and two donkeys. The cattle are for milk, and the goats are for milk, but the sheep are for wool. They have maybe ten percent sheep, and the rest goat. Each winter, before it snows, they begin moving down Swat, towards Chakdara and Malakand area, where their herds will be able to graze. They do not stay in one place only, but move with their herds. They said that the local people down there did not mind, but they had little to do with them. They sold their milk, butter, and ghee in the towns. They did not grow crops up at their home, but gathered plants or bought what they needed. (QN)

3. Inter-valley transhumants: summer

3.1 Shir Muhammad and family group consisting of two brothers and their families were on the road above Chakdara moving towards Dir town. They were coming from a village called Jhang, in a valley just above Chakdara, and moving to another valley above Dir moving to avoid hot weather, which is bad for both people and animals. They are taking their goats and sheep, maybe 200 in total, to graze up in the mountains. The parents will settle in one village up there, and the children, the sons will take the goats further up the mountains to good pastures to graze. They will be up there for about three to four months and they have household things, luggage, pots, utensils all piled onto donkeys to transport. One son remains down in Jhang, to look after the house and to take care of the crops, they have maize and potatoes there. Up in the high valleys they will use ghee and milk to trade to get wheat and flour. (IA)

3.2 Sayyed Rahman, 60-70 years old, a herder in Malana Village, thirty kilometres north of Mingora, in the Usafal Valley. Sayyed Rahman is now too old to travel with his family up to the summer pastures, so remains down in the lower valley at their house. The rest of his family have travelled higher up the mountains with their animals to find grazing for the summer. Sayyed Rahman said he and his family were Gujar, and kept both sheep and goat and cattle. The sheep and goat of his family were around 100, and the cattle around fifteen in number. Every year, after June, all the animals are taken up to the mountains where the family have a hut. The women and the old and young people stay there, while the young men take the animals further up the mountains to find good grazing. The hut is apparently at least one days walk away, and once there, the herd might then move a further three or four hours on. The animals are all milked up on the mountains, and some boys will bring down the milk or ghee to the family, and some for sale. (It was not entirely clear how much contact there was between the actual herders and the family once up on the mountains, but it did seem to be quite regular, at least with regard to the delivery of the milk). Crops are grown down at the house in Malana village, and these are wheat, maize and potatoes. Some vegetables as well, such as okra, chillies and garlic. Wheat is grown over the winter. Each time the seed is scattered and covered from birds, then the fields are left for the snow. The wheat is harvested before leaving in June, dried and winnowed, and then stored in the houses. There is not a lot of wheat, and some is taken up with them to the mountains. Sayyed Rahman said that no-one would take the wheat, or their other possessions left locked up in the house over summer. (QN)

3.3 Sarfaraz Manaras, a summer transhumant. Sarfaraz Manaras and his family are a group of Gujars from Swabi, moving up to spend the summer in Swat. They consider themselves natives of Swabi, that Swabi is their home. They move with buffaloes, every family in the group has one or two cows or buffaloes, and there were a few goats, maybe six or seven in the whole camp. Milk is sold to local people, and milk is the main source of income. They rent houses in Swabi, in exchange for dung and guarding duties. They said they had no goats or sheep (even though at least six were noted) because they are originally plains people. They grow no crops at either the Swabi village, or up in Swat over summer. They buy meat and flour for their own use, and will sometimes sell animals to local butchers. Children tend the animals while grazing, and both men and women work for daily wages. They said that their group was originally from Shanglapur up in the mountains of Swat, but during their grandfather's time, the move to Swabi, and settling there was made. In the fields around their camp, they have constructed the round straw mounds to provide food for the buffalo, because this type of food will give better milk from the animals. The straw was bought from local farmers. (QN)

3.4 Muhammad Roze, has two acres of land about twenty kilometres north of Mingora, in a village called Landekai. He has a herd of around fifty sheep and goat, and two cows, and says he is Gujar. He and his family take all the animals up to the high pasture during the summer to avoid the heat of the lower valley. All the family travel, no-one stays behind. The family stay in a village up there, and then the animals are taken further up to the pastures by boys, and they move around for the whole summer to find grazing. The village they travel to, north of Madyan, is the same every year, and they stay in the same place. Up there, Muhammad Roze and his brothers will work as labourers, and the children will look after other herds for money. They grow some wheat at their home, and potatoes and maize. The wheat is sown when the snow is gone, and they will harvest before winter, when they come back from Madyan. Vegetables are also grown, and these are left over summer. No-one weeds them, those that survive will be harvested. They keep no buffalo, as buffalo are only recent in Swat, and they cannot take them up to the hills. They sell milk and ghee from the goats and cows, both over the winter at home, and over the summer up in the mountains. (QN)

4. Nomadic pastoralists

4.1 Pamu Molu, was interviewed in Lower Dir, while moving with a herd of around 150 sheep and goat. Pamu Molu said he and his family were Ajurs, so that they moved all year round finding grazing for their herds. He said they were from Dir, but did not give the name of a village. In summer he said they took the animals up to the high mountain pasture, then in winter would move down south again, to near Chakdara to graze. Down there was little snow. They never went out of Dir. Pamu Molu said they would stay in an area sometimes for months, and would build huts or shelters for themselves. He said his family had always done this, and the children would do it. The herd would be split for grazing and the children would each look after a part of it. (IA)

4.2 Sham Sahi, was interviewed in Upper Swat near Madyan, when he had come down from the summer pastures in the mountains to attend some sort of formal or official occasion (not made entirely clear quite what). He said he and his family were up in the pastures, where they would move around all summer. They looked after their own animals, maybe 100 sheep and goat, and also the sheep, goat and cattle from villages round Madyan. They would be paid for this, for each animal, and the children of his family would guard the animals, and they had dogs to guard them from leopard(?) or panthers (?) at night. The boys would also carry down ghee, butter and yoghurt for the owners, and for sale. Every day the animals would be milked, and the milk made into butter and yoghurt, which could be carried down. This was only done in summer. In winter, Sham Sahi and his family would care for their own animals, and take them to another side valley, lower down, and then over again, to avoid snow and find grazing, but this was still in Swat. They never grew crops, they ate milk and sometimes meat. They would gather green plants for food, and maybe buy flour with money from milk. (QN)

4.3 Pam Jam, was interviewed while moving with a herd of sheep and goat and some cattle, in Lower Swat. He said his family were ahead, with more animals, and that he had maybe 50 goats, 20 sheep and ten cattle, and that he and his family were Ajur. They had only their own animals with them, but were moving up to Upper Swat, eventually to go to the mountain pastures, and would take other people's animals with them, and look after them for some money. They would be paid more for more animals. Each year they would move up to the high pastures, but not always the same place, it would depend on where the grazing was good. Sometimes the best pastures were several days walk up the mountains. Animals up there would be milked, and children would carry down to the town the butter and yoghurt. They would build a hut, or house up there, and if they ever went back somewhere, could use the same house if they repaired it. (IA)

NB: Sedentary farmers also discussed Ajurs in terms of shepherds for their animals during summer: see interviews 5.1; 5.2; 5.3;

5. Sedentary farmers: Northern Valleys

5.1 Shah Muhammad, a farmer and animal herder owned land south of Mingora in Swat. He owned four or five acres of land, which grew wheat and maize and some rice. He has a herd of 200 sheep and goat, and ten to twelve cows and bulls. Every summer these animals are taken up to the high pastures in Swat. These high pastures are apparently in the mountains above Madyan, moving into upper Swat. They are good for the animals. Shah Muhammad said he employed Ajurs to look after the animals. Shah Muhammad himself is Pathan, and he and his family remained at their home below

Mingora. The animals would be gone from May - June until September, and would be for the winter at the farm. The animals are milked, and ghee and butter is made, and carried down and sold. The land itself has irrigation and terracing, and that is why rice can be grown. Without irrigation from the river, rice is not possible, even thought there is more rain here than Charsadda. Rice is a cash crop, and the wheat and maize and some vegetables are for family use. Over the summer, one cow is kept down for family use. (QN)

5.2 Hazarat Ali, farmer in upper Swat. Hazarat Ali and his family live year round in Peshmal or Pishmell village, Surat district, Kalam. Long ago their predecessors came from Gujranwala in Punjab, and they speak Kohistani, Gujro and Pushtu. They farm, and grow potatoes, garlic, turnips, chillies, maize. They have ten cows and calves, and store dry grass and leaves for the winter. They also buy wheat straw, which is stored inside the house. They stay here all winter, while others may go to Mardan. He said that those who travel do it for the sake of the animals, to ensure better winter food, exercise and so on. Hazarat Ali works for a living, and say that the animals are extra. Because he works, he pays others to take the animals up to the upper valleys to graze in summer. He will pay 100-200 rupees per season, depending on the number of animals. The person, the keeper who is an Ajur, will milk the animals and make ghee, keep some for their own use up in the mountains, and send the rest down to the owner. Hazarat Ali keeps chickens and eggs for their own use, they do not sell any. They are especially important during the winter. The family also has a small beehive on the side of the house. No sheep or goat are kept. During the winter, they can have up to six or eight foot of snow, so they need to always keep the path from the house to the road clear. The government provides a bulldozer to clear the road to the village. (QN)

5.3 Rozi Gul, a year round sedentary farmer in upper Swat. Rozi Gul and his family live in Charat Village, which is north of Kalam, in upper Swat. Rozi Gul is a Swat - Kalam Kohistani. (This is the answer he gave to questions about his ethnic and social group. The interpreter suggested he was Gujar, but he maintained he was Kohistani). He is a farmer, cultivating potato and maize, which is the most important crop, and also the tree cops of almond and apple. He also has sheep and goat, and since about eighty years back, cows. His family started farming to feed the animals, and before that they were permanent travellers. They settled on this farm maybe twenty years ago. Rozi Gul's predecessors were seasonal, and did move with animals. They would spend six months down in Mardan, Swabi, Charsadda area. When he came to Kalam to settle, he started farming, and began keeping cows and bulls. Now, his family do not move seasonally. Instead, they store dry grasses and leaves of corn, and also purchase wheat straw to keep their animals fed over winter. Storage of animal food is done in two places: one a sort of lean-to on a fence, and also in a room in the old house, right beside the animal shelter. Of his predecessors in Dir and Chitral who travelled, only some would go, but they would take all the animals. Some would stay to look after the house, the old people and so on. During the summer, they would return to the same place, their home up in the valley. They had no crops, but would store bought crops to see them through the winter. They would sell dried fruit and wool products, and animals and animal products. Crops are now important, mainly to ensure food for animals, but back then, when they travelled, it was only animals that were important. As this area becomes completely snow-bound, during the winter, they have to stay inside for four months every year. They have to have a fire inside the house, and the animals are also kept inside for four months. During September they will start storing dried grass. The yearly birth rate is for sheep, one lamb every twelve months; cows, one calf every twelve months; and goats, one kid every six months. Buffalo are kept only very occasionally. If farmers in this region have more than fifteen sheep or cows, and more than twenty to thirty goats, then they are sent to the upper valleys for four to five months for food, in the care of a shepherd, or Ajur, then returned for winter. The milk and ghee and other products are brought back down to the owners for use or sale, and the Ajurs reserve a percentage for themselves in payment. Only those with enough animals can send them up, if only a few animals are owned, they must be kept down for their own milk through the summer. Dung is stored for both fertiliser, and dung cakes are stored for winter fuel. (QN)

5.4 Abdul Raziq, herder in Upper Swat. Abdul Raziq lives the year round in Tandachena Village, Chil Valley, Madyan, Swat. He has goats, cows, sheep grazing on the upper slopes of the valleys, some three hours walk away. Abdul Raziq said he and his family were Gujar. There are around fifty sheep and goat, and around ten cows. The animals are cared for by young males in the family, and they bring down milk and ghee for sale. The whole family also lives here in the village all through the winter, feeding his animals on straw and grass, dried in the summer. Again, the animal quarters were directly below and accessible from the house. Abdul Raziq has no crops at all, and neither he nor his family have ever been farmers. This seemed to be because keeping animals and selling their products was more cost effective. (Although this was not able to be directly confirmed in interview, and may be due to social or ethnic constraints). In the old days, his grandfather's time, they had more sheep, goat and cow than these days, because they were living "on other side" of valley, but now they own their own house. (The importance of settling down seemed to be that this restricted their movements, and so access to grazing, rather than any loss of animals as capital in exchange for property). This area is covered in snow in the winter, but it is still possible to get to the village, they keep the paths by the irrigation channels open. Because they can get to the village for food, they only need to store animal food, and can take milk in to sell and buy what they need for themselves. (QN)

(NB: after this interview, we talked to Pathans in the village, who told us that this family were Ajurs, who had kept animals, and had no settled house. However they have now settled, and so are calling themselves Gujars.)

5.5 Taj Mahummad Khan, a year round sedentary farmer in mid Swat. Taj Mahummad Khan lives in the village of Kokari, in the Jambil Valley and he described himself as a Gujar, who now was a farmer. He has lived in the same village for nearly thirty years, and before this time, he was still in the village, but farmed for someone else. He farms the land and guards it, and gives the farming produce to the owners of the land, they decide what to grow. In return, he can keep animals on the land and keep any profit from the animals or the animal products. Taj Muhammad Khan has three buffalo and three calves. He also has one goat, he used to have a flock of thirty to forty goats, but these were sold for money. He has his own chickens, and sells both chickens and eggs for money. On his farm, he grows wheat, and the grain is for humans, but the straw and leaves are all used for animals, Trees, however, are the main crop: there are more than one thousand tress, mixed orange, persimmon, pear, plum, apple, walnut. Buffaloes are kept in the building, room adjacent to the house, and the goat in a room next to the buffalo. There is also one donkey for carrying wood. The division was described as the landowners owning the product of the land (with the exception of the 'waste' such as straw, and some produce kept for the family needs) and the farmer owns the animals and the animal products. In winter, there are three to four months of cold and snow, but for those who live closer to a village, things are not as bad these days as they were previously, in grandparents time. (QN)

5.6 Pai Mahummad, a year round, sedentary farmer in mid Swat. Pai Mahummad works a small farm in Luk Maira village, on the same road as Kalako-deray, slightly further north east along main Jambil Valley road. Pai Mahummad says he is a Gujar, but he now lives on a very small area of land, one or two acres, that is owned by a farmer. He grows and sells maize and tomatoes, and produces straw for animal fodder, which is dried and stored in the fields for winter. Pai Mahummad has five cows, the milk from these cows is sold, and the cows spend winter in a ghwajal, or room, adjacent to the house. In winter, snow is common, and the cows must be fed straw, which he sometimes has to buy. During the summer, the cows are grazed on any open land, including along the roadside, taking advantage of grass, herbs, shrubs, and are kept outside most of the time. The exception to this is the calves, who are kept inside for at least part of the day. Chickens are also kept, he was unsure how many he had, but chickens and their eggs can be sold, as well as used for the family. No sheep and goat were kept, selling milk from the cows was a better way of getting money. Pai Mahummad also has a nursery of mountain plants further up the valley. These plants are bought by the government for the Forestry Department, who use them for re-planting de-forested areas. Grain is bought, and is very expensive, therefore, it is stored within the house, in a special wooden chest. Nothing is stored underground, pits are not known. Animal food, straw is stored in the fields, covered by plastic, and surrounded by thorn branches. Human food is stored within the house. (QN)

5.7 Zahir Shah, a farmer at the site of the Shahnaisha Buddhist Stupa, outside Saidhu Sahrif, Swat. Zahir Shah said he and his family are Gujars, originally from Ragastan, north of Kalam, but their families migrated from Kalam to lower Swat, and settled here. He and his family are now both farmers, and caretakers of the Shahnaisha Stupa. The animals he said they kept were four to five buffaloes and one or two cows, for milk, and then a zebu bull for ploughing. Zahir Shah said they had no sheep or goats (however, two kids eating branches that had been collected by children were noted during the interview). The crops grown include okra, tomato, potato, garlic, maize and wheat. Animal food is grass that has been collected specially from hill top areas, brought down to the house, dried and stored for winter food. This is done in the summer, and the grass is dried in the sun. Wheat straw is also collected for use as animal food, and stored in round stacks outside the houses, covered in plastic sheeting. Zahir Shah said that while he and his family consider themselves Gujars, and so pastoralists, have become agriculturists, farmers, also, because they could no longer promote their own professions of caring for buffalo and cow, so needed another way of gaining food for animals and humans. His father is over 80, and he was born here on the farm, so farming was begun some time before that. In winter, animals are kept in a separate room adjacent to the house, called a ghwajal, where they are fed the straw and dried grasses. Animal dung from the ghwajal is collected and spread on the fields. Pits are not dug - no storage in pits is ever used. This is due to the possibility of landslides, although forests and trees may reduce this risk, but pits are not known. (QN)

5.8 Dilawai Shah Husain, a farmer in Nalan Village, Chil Valley, west of Madyan, Upper Swat. Dilawai Shah Husain has land of around five acres in the village of Nalan, which supports his family of fifteen people. On the land he grows wheat and maize, and some vegetables. He also fruit trees that are grown two to three hours walk up the valley, in the mountains. For animals, he keeps sheep and goat and some cows, and during the summer these are taken up the valley to the high pastures. They are taken by men of the family, and kept up there for four months. The animals are milked, and the butter and ghee sold in towns. During the winter there is always snow, and the animals are kept under the house in a shelter. (This shelter was directly accessible from the house). The animals are fed dried grass and straw, and the dung collected for fuel, and later fertiliser. The fruit trees were apple and walnut and mulberry, and the fruit was sold dried and fresh, some was kept for winter for food. (QN)

6. Sedentary farmers: Bala Hisar

6.1 Ahmadh Ali Shah farmer, son and grandson of farmers, from the village of Mani Kheki, five kilometres north east of the Bala Hisar of Charsadda. There are four seasons per year and two crops, with vegetables possibly a third. Maize is also grown as a third crop. Sugar cane is planted in March (grows for nine-twelve months) and is harvested in November-February. It is sown as small pieces in lines, and the plants are kept for three years, then removed. The farmer will stagger crops and mix sugar cane crops with maize, wheat and animal fodder grass. The farm size for one man and his family, comprising eight to ten people, would be ten hectares, which will easily feed them. In the village, there are two parts divided by the main irrigation drain. One side is government irrigated and has lesser amounts of water for lesser amounts of time, and is often insufficient, so extra is requested from the non-government side. On the non-government side the old drainage has been cut by local people, the source is in the Malakand Hills, so impossible to cut out upper parts and farmers. The government side utilises water from the upper Swat Dam. In Dir, Swat and Chitral, can grow rice close to rivers, and with floods can grow crops such as wheat. Floods up in the valleys help fertility also. The Upper Swat canal has meant that previously barren, waste areas abounded in among fields and houses, and his Grandfather used to wander over them. Now these waste areas are sunken fields. His grandfather's flock was maybe 280-290, mostly cows, big oxen and zebu, sheep or goats. Milk was supplied to towns on donkeys in big ceramic vessels. Flocks would travel eight to fifteen kilometres around the village in a year, with one to two people, maybe from family, maybe paid to look after them. Maybe there would be the flocks from a number of families. Others left in the village would cultivate; in pre-irrigation times, dry wheat only, for one season per year. Sugar cane was found in low, swampy areas only pre-irrigation. Ahmadh Ali Shah noted a diet change from his Grandfather's time to present. Most people used to mainly eat milk, butter, cheese, lassi, bread, maize, wheat, rice, and barley. Vegetable is later, more modern, and meat was also less important in the past. Sugar cane is mainly a cash crop, but also grown for selves. If dry in field, used as a fuel and animal fodder. With regard to harvesting, wheat, the main food crop is sheafed after cutting. Ahmadh Ali Shah also talked about people coming from the hills to his village. They come from Kalam in Upper Swat and bring herds of cow, sheep, goat. They are given a free house in, around the village, and mean that there is a labour supply for the harvests of sugar cane and wheat. The house is permanently available, but to them (the incomers) it is temporary. They are paid for work and are a good labour source. During the harvest, they cut sugar cane, clean cane, pack it, process it using the machines to make gur, and pack this also. The top of the sugar cane, described as 'grass', is given to animals as fodder. The children often look after flocks, and taking them to suitable grazing land, such as government owned land, or in villages where they can graze on open land. The animals can also graze fields where the crop has been cut. Sometimes these groups will bring dried fruits, such as almonds or mulberries down as gifts, and take gur back. They generally return maybe April, or May, after the wheat harvest. These groups now often move by truck, some by foot still. Dogs are important as guards. These people have fixed areas that they have travelled to for generations: to stay, work, leave. They have their own land in the mountain villages, where they are both agriculturists and pastoralists. Seeds can be sown, scattered, and covered by snow - then they leave, very few people stay up there - those that do need to prepare fuel, food, and for animals to survive. They don't wash very often. Now there is central aid in most mosques. They arrive in the village in the Vale of Peshawar at the beginning of November, and the time of journey may take four to ten days. (IA)

6.2 Lal Bad Shah the farmer, but not the owner of this land. He is a share-worker, and shares the produce on a fifty-fifty basis with the owner. This is apparently very common around Charsadda, where the owner of the land may teach, or have a similar occupation, and live in the towns. There are fourteen acres in this one farm, so it is quite large for this area. La Bad Shah said he was Pathan. Lal Bad Shah prepares his land for planting sugar cane following the wheat harvest using a wooden plough with iron blade attached, and two zebu pull the plough. The two zebu are used for ploughing, also pulling carts and transporting crops. Two cows are used for milk, and producing calves to sell. He also has one buffalo for milk, for his own family. On the land Lal Bad Shah grows a type of vegetable, (*Amaranthus* sp.), a leafy green vegetable similar to spinach as well as wheat and sugar cane. The vegetable is available all year round and grows in with the other crops. It is not specifically cultivated, is gathered. Can be dried on top of houses, stored, and even sold commercially, and is considered a supplementary food. Two crops of wheat are grown in a year, and the wheat is very well weeded and tended. Flood irrigation is used, with channels extending all round the Bala Hisar and surrounding fields. The channels are used both as property markers and to irrigate. Flooding occurs in July/August always. Sometimes there are winter rains in November, December, January. In the village an announcement is made, and a communal decision is reached for when and where to irrigate, based on the amount of land each farmer has. The farmer will block the river five kilometres away and divert it into the series of small channels. In summer there is weekly irrigation, in winter there is monthly irrigation. Mixed cropping is common, where sometimes wheat and sugar cane are grown together, sometimes sugar cane and animal fodder grass, a type of clover. Sugar cane cleaning is carried out in a circular packed earth area, and two oxen are driven round to thresh and crush. The sugar is then cooked and concentrated and a deep pan is used to collect the juice. When extracted into a pot, the gur, or raw sugar, is then sold as a cash crop. Cane waste, the straw, is used as animal food. Sugar cane needs much water to grow. (MD)

6.3 Rahman Said, a farmer originally from Dir, around 60 years old. Rahman Said has lived permanently in Charsadda for fifteen or sixteen years. Here he has tow or three buffaloes, some land and trees. In Dir he had a low level of income, he was a share farmer growing wheat and maize, but no sugar cane. Wheat was a summer crop, and maize was a winter crop. He also grew some potatoes. In Dir he had sheep and goat, suited to the hills, no buffalo, cow or oxen. There are small cows for the hilly areas, bigger, heavier animals for the forest. Now he has land to work, and he grows wheat and some sugar cane. Sugar cane is a cash crop, and cannot grow up in Dir. He does not own his land, but is still better off here than in Dir. He can grow wheat twice per year, and sugar cane once. He has no sheep and goat down here, only the buffalo, which are good for milk and ploughing. Rahman Said says when he was young, he and his family came down every year from Dir with their flocks of sheep and goat, to this place by Charsadda, the same place very year, and that is how he knew that he would be better off down here with his family. Most of his family (brothers) remain up in Dir, where they have little land and no buffalo. His wife and children are now here. He said his Grandfather was Gujar, but he is not. (He did not describe himself as any particular ethnic group, although this may have been as much to do with not being able to ask this question in the right way, or the presence of the interpreter, as a real absence of ethnic affiliation). (MD)

6.4 Rahman Said, farmer on land adjacent to the Bala Hisar. Rahman Said is a share farmer, of only four acres. He has one buffalo, one bull and shares both bull, buffalo, and oxen with other farmers around him. The owner directs him to grow sugar cane as a cash crop. Wheat for his own use is harvested in May, then the land is prepared and ploughed, and a new crop sown in September/October. He and his sons weed the crop and carry out the other tasks necessary. The sugar cane is planted in march, and harvested around November. They use their oxen or bull to thresh, and will mix their crop with other farmers to prepare it for sale. Men who have come down for the winter might help with the sugar cane work, in exchange for wages. Soil from the site of Bala Hisar is used as fertiliser on fields for a sort of topping up of the land. The mound soil is fresh, unused for many years, whereas their own soil is now overworked and stale. No sheep and goat are kept, they are not needed. Sometimes, sheep will be bought at market for eating, but mainly the family have milk and flour, and they buy some lentils and rice. Chickens are kept too, for eggs, and meal if guests come. Women look after chickens at home. Some vegetables are grown in the compound, and these are looked after by sons and their families. (MD)

6.5 Mr Said, 65, on land adjacent to the Bala Hisar. Mr Said is a share farmer, sharing the products of the land fifty-fifty with the owner. He farms three acres, which is not sufficient for the whole family (five men and their families) so three work outside land also, for other farmers and in town. The owner decides what is grown, and always has. He has worked these same three acres for twenty years, and he grows only sugar cane, which is a full year crop. It is planted in January/February, and harvested in late November/December. The sugar canes are chopped, and the pieces planted for the next years crop. All the crop is sold in Charsadda, and Mr Said processes the sugar himself locally, using his bull for the threshing, and his sons for labour. He does not grow wheat, or clover, but does grow some vegetables in a separate area (near the house). He also has one bull to help process sugar, and one buffalo to milk, but no sheep or goat. Chickens are also kept. (MD)

6.6 Mr Mustaqim, farmer on land adjacent to the Bala Hisar. Mr Mustaqim is the owner of four acres which have come down from his great grandfather. Eight people are all supported from the land, that is him, his wife and four small children, and his mother and father. The crops that he grows are sugar cane for cash, and wheat in summer, a very small amount, for the use of his family. Wheat is planted in February, harvested in May. After the harvest it is sun dried, and the straw is given to animals, and the grain is sent to the local Charsadda mill for grinding to flour for the family. For many years he has grown this, and plans to grow same for years yet. He has one cow for milk, one bull for traction, some chickens, but no sheep or goats. He said he had no need for sheep or goat. Irrigation is necessary for sugar cane, and he pays to the government between 80-120 rupees for the whole year for water once a week, more or less half an hour per one acre. (MD)

6.7 Baghi Shah, a farmer on land adjacent to the Bala Hisar. Baghi Shah is a farmer with a half-half share of produce with the owner of the land. He farms six acres for eleven people, and they all work the land (although again it was not clear whether the total family comprised eleven, or whether this was the male number). Baghi Shah and his family are Pathans. Baghi Shah grows some wheat, very small amount for family only, the rest of the land is all planted in sugar cane, all cash, and the land needs to be irrigated once a week. The two crops of wheat and sugar are grown together. It is the land owner rather than the farmer who decides what is to be grown, and this is the same every year. They have one cow for milk, two bulls for traction and sugar cane processing, and ploughing, which happens every two - two and a half months. They do not keep any sheep or goat, but buy this meat at market. Some chickens are kept. Wheat straw is given to cattle, and they need to buy good fodder for the cow to give good milk. The food for the bulls is not so important, because they are not milked. (MD)

6.8 Zair Ullah Khan, farmer on land adjacent to the Bala Hisar. For the last twenty years, Zair Ullah Khan has worked the same land, a farm of five and a half acres. The land is owned by someone else, and the money made is shared fifty-fifty with the owner. This farm is supporting twenty to twenty-five people. They only ever grow wheat for family use and sugar cane for cash, and the owner of the land decides what is grown. Irrigation, from the government, is one half hour per acre per week. Sugar cane is only source of income. From one acre of sugar cane, they can earn 40,000 rupees per year (half to owner), but this depends on the quality of the land, and rainfall. Weeding is done every three months for sugar cane, no weeding for wheat. Irrigation by channels leads to silt build up on the fields. This silt can be left, and then eventually after four to six years a hired tractor will be brought in to clear it, and lower the fields again. Otherwise the silt will make the fields too high to be reached by the irrigation water. Zair Ullah Khan has one bull for traction and ploughing, one buffalo for milk., and the animals provide about half the fertiliser used, half is market bought. If he and his family want meat, they will buy sheep or goat, but for his own needs, he has only big animals. Zair Ullah Khan has some contact with transhumant groups, mainly employing them as labourers to process his sugar cane in the sugar mill. However, he said that they were the same people who came here, and grazed their animals around his farm, and he knew them. (MD)

6.9 Zair Ullah, farmer on land adjacent to the Bala Hisar. Zair Ullah farms the land and shares profit half/half with owner, who decides what is grown. He works twelve acres for fifteen people, they work only on the land. He has had the same land for 50, 60 years, and it was worked by his father and grandfather before him. He keeps two bulls and four cows, but grows no animal fodder, his animals are grazed on common or government land. He has no sheep or goat, but does have chickens, which are good for eggs and meat. He grows wheat and sugar cane, that is five acres of wheat with sugar cane, and seven acres only sugar cane. Mainly his family provide labour, or sometimes they will share with neighbouring farmers. He knows about those who come from the valleys each year, but has little contact with them. Sometimes these men might help out with the sugar cane harvest in exchange for wages. (MD)

6.10 Said Afzal, farmer, adjacent to the Bala Hisar. Said Afzal has worked this land for sixty years, it is owned by a man who lives in Charsadda town. Only wheat and sugar cane is grown, and he can only ever remember these being grown. The owner decides what is grown, and he wants sugar cane to bring in money. The wheat is for Said Azfal's family use. The farm has five acres, and there are eight people who live from it. He has two bulls and one cow, but no sheep or goats. The cow is for milk for his family, and the bulls are for ploughing, twice a year and for the sugar cane processing. Only the cow gets fodder, or food bought, the bulls are grazed around the area, and tended by children from his family. Labourers are hired in to process the sugar cane, he pays in money and food. These labourers can be from Swat or Dir, there are always people from the valleys in the area when it is time to process the sugar cane. The sugar cane leaves and wheat straw are given to the cattle, some are dried and stored for later use. (MD)

6.11 Usafal Khan, a farmer on land adjacent to the Bala Hisar. Usafal Khan works a farm for fifty-fifty share with owner, and this is a larger farm, of ten or more acres, so fodder crop is also grown. They can get 8,000-10,000 rupees per one acre for animal fodder, or 4,000-4,500 rupees per quarter acre, but sugar cane will give more money. The advantage of fodder, is that it can be cut four times a year, and will re-grow like grass. Only those with more land will grow fodder. It is sold in towns to those with animals. Two buffalo, three cows and two bulls are kept by Usafal Khan, to provide milk and ghee for his family, and to plough the land. Only enough wheat is grown for his family to use, the rest of the land is used for sugar cane and fodder, decided by the owner, to bring in money. They buy rice and lentils for the family to eat, no rice is grown at Charsadda, only up in Dir and Swat. Rice needs irrigation, and there is only water here for sugar cane. They grow some vegetables in the compound. He has no need to keep sheep or goat, but he does keep chickens. The manure from his bulls and cow is used on the land, but still some is bought in, or soil from the Bala Hisar itself is used. This soil is fresh and good for the land. Usafal Khan has on his land the house in which Shah Nazar (see interview 1.4 above) and his family stay during winter. He says they have always been coming, and in return for the house, he gets dung and the men help with the harvests. The land owner does not mind them staying there. (MD)

6.12 Waheed Ullah, a farmer on land adjacent to the Bala Hisar. Waheed Ullah farms land that is owned by someone else, and the money is shared half/half with owner. This farm has eight acres for five people. Half of his acres are in fodder crop, and he also grows sugar cane, but no wheat. The sugar cane is mixed with the fodder, and the fodder is for his animals only, only the sugar cane is sold. This farmer decides what to grow, rather than the land owner, but as far as he remembers, his fore-fathers grew the same crops. He has two bulls, but no sheep or goat. Some chickens are kept, and are eaten as well as eggs. When sheep or goat are needed, they buy them from the market. He employs three labourers who work for whoever asks them, and also for harvesting of sugar cane, he employs those here for the winter. Waheed Ullah says the same people come year after year. Like those who work the land here, the same happens for generations. (MD)

www.ingramcontent.com/pod-product-compliance
Lightning Source LLC
Chambersburg PA
CBHW041706290426
44108CB00027B/2867